Real-Time Massive Model Rendering

Synthesis Lectures on Computer Graphics and Animation

Editor
Brian A. Barsky, *University of California, Berkeley*

Real-Time Massive Model Rendering

Sung-eui Yoon, Enrico Gobbetti, David Kasik, and Dinesh Manocha

ISBN: 978-3-031-79530-5   paperback
ISBN: 978-3-031-79531-2   ebook

DOI 10.1007/978-3-031-79531-2

A Publication in the Springer series
SYNTHESIS LECTURES ON COMPUTER GRAPHICS AND ANIMATION

Lecture #7
Series Editor: Brian A. Barsky, University of California, Berkeley

Series ISSN
Synthesis Lectures on Computer Graphics and Animation
Print 1933-8996   Electronic 1933-9003

Real-Time Massive Model Rendering

Sung-eui Yoon
KAIST

Enrico Gobbetti
CRS4

David Kasik
Boeing

Dinesh Manocha
University of North Carolina, Chapel Hill

SYNTHESIS LECTURES ON COMPUTER GRAPHICS AND ANIMATION #7

ABSTRACT

Interactive display and visualization of large geometric and textured models is becoming a fundamental capability. There are numerous application areas, including games, movies, CAD, virtual prototyping, and scientific visualization. One of observations about geometric models used in interactive applications is that their model complexity continues to increase because of fundamental advances in 3D modeling, simulation, and data capture technologies.

As computing power increases, users take advantage of the algorithmic advances and generate even more complex models and data sets. Therefore, there are many cases where we are required to visualize massive models that consist of hundreds of millions of triangles and, even, billions of triangles. However, interactive visualization and handling of such massive models still remains a challenge in computer graphics and visualization. In this monograph we discuss various techniques that enable interactive visualization of massive models.

These techniques include visibility computation, simplification, levels-of-detail, and cache-coherent data management. We believe that the combinations of these techniques can make it possible to interactively visualize massive models in commodity hardware.

KEYWORDS

Visibility computation, Simplification, multi-resolution rendering, levels-of-detail (LOD), cache-coherent data management, Massive model rendering, interactive visualization and rendering, Large-scale rendering, Rasterization, Ray tracing, Collision detection, cache-coherent layouts, data explosion, Occlusion culling, Large triangle meshes, View-dependent rendering

Contents

CHAPTER 1

Introduction

Interactively displaying and visualizing large amounts of data has been a challenge in computer graphics since its inception. In many cases, the amount of data that a user wants to visualize exceeds available processing power and memory capacity. Digital computers have natural limits dictated by physics, mathematics, and cost considerations. Interactive performance, which forces the computation of new frames at 10 Hz or faster, exacerbates the problem. Making a system 'interactive' means that any solution must address real-time performance and cognitive, perceptual issues.

Physical limits have increased dramatically since Charles Babbage developed the notion of a programmable computer in the mid 1800's. The cognitive issues have stayed consistent for hundreds of years. In computing, 'exceeding scale' generally means that some constrained system resource becomes saturated or overloaded. The resources include raw processing power, display processing power, memory size, disk size, network capacity, and preset size constraints.

The net result is that these constraints impose limits on what users can expect in terms of overall capacity, performance, and capability of real-time, interactive applications. It has been our experience that users invariably expect more than computers and computer scientists can deliver. Users always want to gain more digital insight and exceed computing system limits.

This is certainly the case in interactive visualization. Numerous domains create highly complex digital models. Examples include: industrial CAD models of airplanes, ships, production plants, and buildings; geographic information systems; oil and gas exploration; medical imaging; scanned 3D models; un-organized information spaces; and high-end scientific simulations. The digital models may contain millions, even billions, of 3D primitives. The primitives include points, surfaces, voxels, and higher-dimensional data forms. Each primitive is often associated with a complex set of parameters. We can store the data, post inquiries to search engines to analyze it, produce reports (including still pictures and films), and derive other information about it. We have just not been able to see it in real time.

In this chapter, we provide an introduction to key algorithms for massive model visualization. The dictionary [Ame07] defines massive as:

1. Consisting of or making up a large mass; bulky, heavy, and solid.

2. Large or imposing, as in quantity, scope, degree, intensity, or scale.

3. Large in comparison with the usual amount.

The models addressed here are massive in all three senses. The digital datasets representing the models describe high levels of detail that may not be visible to the human eye until magnified. The data can consume tens of gigabytes and even terabytes of storage, a billion or more geometric primitives, and range in units from light years to angstroms. And, the data exceeds the usual capacity of conventional processing techniques.

Figures 1.1, 1.2, and 1.3 provide three samples of massive models.

Massive model visualization seeks to provide users with the ability to interact with 3D models of almost unlimited size and complexity—mainly with respect to geometry but increasingly in terms of appearance, illumination, visibility, and other features that create the illusion of photorealism.

Interactive performance involves generating new frames quickly enough to convince a person's visual system that movement is continuous. Developing a solution achieving a sustained and consistent perfor-

Figure 1.1: Landscape with over 1 billion polygons (courtesy Saarland University).

mance level requires a system-level solution. Like many performance-tuning efforts, addressing only one or two parts of the system may cause other aspects to fail. Moreover, some approaches we discuss handle one class of massive models well and are unable to achieve the real-time, interactive performance for different classes.

1.1   BRIEF BACKGROUND

The problem of rendering complex models at interactive rates has been studied in computer graphics and related areas since Ivan Sutherland developed Sketchpad in the early 1960's [Sut63]. The attack has been across-the-board, and researchers have developed new mathematical representations, data structures, software algorithms, and hardware designs to cope with scale.

A full review of the history of the field is beyond the scope of this book. Examples of the types of innovation include:

- **Level of detail:** Jim Clark [Cla76] proposed the idea of hierarchical representations and use of multiple levels-of-detail of objects or scenes to accelerate the rendering.

- **Culling:** Trivially rejecting objects not in a viewing space decreases the overall workload. Culling types reject 3D data based on the view frustum, back faces, sub-pixel coverage, and occlusion. The University of North Carolina at Chapel Hill (UNC) Walkthrough Group and the University of California Walkthrough Group developed several of these concepts; for example, see [ZMHH97, BSGM02, ACW+99, FKST96]. kd-trees [Moo91] and other space partitioning techniques are often used to cut down the candidate visible geometry as an adjunct to culling.

Figure 1.2: Double Eagle Tanker with 82M Triangles (image courtesy University of North Carolina; model courtesy of Newport News Shipbuilding).

- **Showing highly simplified geometric shapes when performance drops:** One of the first industrial systems to render large CAD models was the Boeing FlyThru system [AM96]. FlyThru uses simple boxes to improve performance. FlyThru could only display as much data as could fit in a workstation's memory, about 1/50th of a complete Boeing 777.

- **Memory management:** The notion of working from disk as the source of geometric data starts appearing in the mid-2000's. [Yoo05] and [Bru07] are good examples of memory management schemes for disk-bound processes.

Work on massive models visualization continues to focus on both GPU-based rasterization and interactive ray tracing rendering approaches. Both use parallel processing extensively, although in dramatically different ways. Modern GPUs embed hundreds of fragment processors and use them in parallel for rasterization. Examples of GPU-based solutions include North Carolina's Walkthrough system [Bro92, ACW+99] and Ilmenau's Interviews3D [Bru07]. Over the last seven to eight years, real-time ray tracing has emerged as an alternate method. Even though ray tracing has been extensively studied in computer graphics for more than three decades, general purpose CPUs were not fast enough for interactive ray tracing. The renewed interest has been spurred by the exponential growth rate of processing power and hardware trends of using multiple cores. Since ray tracing algorithms are embarrassingly parallel and easily map to multi-core and multi-processor systems, it is expected that the performance of ray tracing continues to improve significantly. Efforts are underway to develop faster algorithms that utilize the SIMD capabilities and the multiple cores of upcoming commodity processors [DSW07].

Figure 1.3: Power plant with shadows (image courtesy University of North Carolina).

1.1.1   Organization

The rest of this chapter provides motivation for the concept of massive model visualization from a user's perspective, the most common types of application datasets, and key system implementation issues. The next four chapters provide an overview of specific components needed for achieving real-time performance. As progress is made in these areas, other performance bottlenecks will occur and must be addressed to achieve a complete system suitable for production use.

The performance bottlenecks covered in this monograph involve finding creative solutions for:

- Converting large amounts of 3D data into properly colored pixels on a screen to determine what is visible and not visible (Ch. 2).

- Reducing the complexity of the data that must be processed on a frame-by-frame basis by using adaptive mesh simplification techniques (Ch. 3) or alternative representations (Ch. 4).

- Moving the sheer amount of data involved in massive models from secondary storage through the memory hierarchy on general purpose and special purpose processors. Techniques to create fast storage-to-memory methods require development of cache-coherent schemes (Ch. 5).

In addition to the three above areas, there are other critical aspects needed to achieve a complete system level solution for massive model visualization that are not covered in this monograph. The other areas include:

- Data acquisition and modeling methods.

- Data marshaling and configuration management.

- Dealing with continually changing and time-dependent models.

- Data preparation (pre-processing).

- Programming techniques for multi-processing, multi-threading, and multi-core hardware.

- Special purpose hardware dedicated to interactive visualization.

- Distribution strategies for massive amounts of data.

- Data quality and data interoperability.

Applying a successful approach to other datasets extends the system problem specification to finding, assembling, and updating the massive amount of data itself. This means that data configuration management is essential. The notion of data configuration management assures that users actually see the 'right' data. Entire disciplines (e.g., product data management, document management) are devoted to the art of configuration management. [Ste07] provides a view of CAD data configuration management for visual data at General Motors. Similar issues are pervasive throughout all domains that generate massive amounts of data and complement massive model visualization.

If nothing else, the reader should be convinced that solutions to the problem cannot be isolated to an application, a rendering approach, a modeling methodology, a network speed, or a delivery system design improvement. Real-time interaction with complex models truly requires a system solution.

1.2 MOTIVATION

The simplest way to characterize the reason for pursuing massive model visualization is that people are getting overwhelmed by data. Somehow, we have to be able to use our cognitive abilities to transform the data into information that can eventually be used to make decisions, improve products, increase understanding, etc.

The visual channel features the broadest path into the brain, and this is why we emphasize interactive, graphical techniques. As the size of the data grows, the size of corresponding visual representation also tends to grow. At this point, the human visual system does not get saturated as quickly as the computer graphics techniques do. Significant amounts of research have been done in this area, and the classic book on the topic is [Gib50].

1.2.1 Data Explosion
Most readers are well aware of the growth of data in their daily lives. The digital data sources seem to expand daily and range from cell phone to Blackberry to e-mail to YouTube to iTunes to actual work tasks.

[LV03] documents the last year of a multi-year project to estimate the amount of digital data stored and transmitted on a yearly basis. The project team concluded that:

- Print, film, magnetic, and optical storage media produced about 5 exabytes of new information in 2002. 92% was stored on magnetic media, mostly in hard disks.

- This amount of new information is about double of the amount stored in 1999.

- Information flows through electronic channels (telephone, radio, TV, and the Internet) contained 18 exabytes of new information in 2002. This is 3 1/2 times more than is stored. 98% is voice and data sent telephonically via fixed lines and wireless.

Figure 1.4 summarizes just what the values like "exabytes" mean.

Name	Value	Equivalent
Kilobyte (KB)	1,000 bytes, 10^3	Typewritten page
Megabyte (MB)	1,000,000 bytes, 10^6	Small novel
Gigabyte (GB)	1,000,000,000 bytes, 10^9	Pickup truck filled with books
Terabyte (TB)	1,000,000,000,000 bytes, 10^{12}	2 TB: academic research library
Petabyte (PB)	1,000,000,000,000,000 bytes, 10^{15}	200 PB: All printed material
Exabyte (EB)	1,000,000,000,000,000,000 bytes, 10^{18}	5 EB: All words ever spoken

Figure 1.4: Giving meaning to exabytes.

Some people, like intelligence analysts, are working to gain a better understanding of the totality of as much of this data as possible. The idea is that analyzing more data is better than analyzing less. They are using visual techniques to complement more traditional analysis methods [TC05]. The community focuses not only on visual analytic methods but also on the graphic communications techniques essential to present results in a way that others can understand.

Much of the work we do everyday relies on search tools (e.g., Google, Yahoo). Commodity search engines assume that data is textual, and the data types involved in massive model visualization do not lend themselves to text search. Furthermore, visual context is often essential when trying to understand a dataset and how it fits together, both internally and externally, with other datasets. Complex visual datasets like Google Earth and the Visible Human ([oM95]) have become the norm in their respective domains.

1.2.2   Human Vision and Visual Analysis Tasks

Vision gives humans the broadest and most flexible input channel to the brain. People have a remarkable ability to assess a complex scene quickly and focus on its most salient features. Computer-based visualization tools attempt to cope with the continual increase in CPU power Gordon Moore originally predicted in 1965 [Moo65] as balanced against the increase in human brain processing power, also called God's law. Figure 1.5 contrasts Moore's law and God's law. [Bux02] developed this conundrum in terms of users interacting with computer systems, and it applies equally well to visualization.

Massive model visualization tools provide users new capabilities. The capabilities have a number of different uses because people can perform a number of tasks based on visual analysis alone. [Kas04] defines typical visual analysis techniques in the aerospace industry. Visual analysis lets people:

• Find an object in a complex scene given:

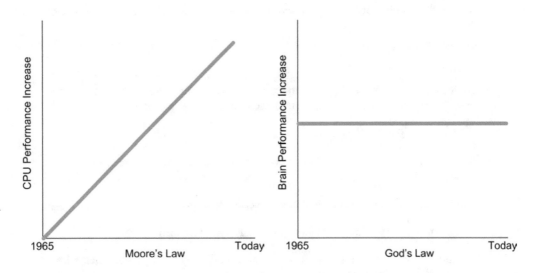

Figure 1.5: Contrast between Moore's law and God's law.

- The physical object.
- A picture of the object.
- A mental image of the object.
- A verbal description (e.g., 'something that looks kind-of-like').

• Focus on the found object to better understand surface characteristics (e.g., smoothness, roughness). The characteristics can be determined by direct visual inspection or by understanding the results derived from physical characteristics (e.g., aerodynamics flow, stress).

• Once the object is found, look at objects in the immediately surrounding volume to:

- Identify discrepancies in space consumption (do the pieces occupy too much of the same space?).
- Determine interference/overlap by direct visual inspection or by understanding the results derived from computing overlaps. In either case, visual analysis is necessary to determine if the interference is acceptable.
- Find gaps/voids to determine the proper clearance between/among objects. Visual gap analysis is often supplemented by a method to measure distance between objects.
- Trace a path to find the other end of connected objects. In aerospace, such connectors are most often 'long, skinny things' like hydraulic tubes and wire bundles. See Fig. 1.8 for an example.

• Visually scan the scene to discover:

- Misplaced objects (homage to hanging chad).
- Forgotten objects ('Drat, we forgot a wing!')
- Patterns from objects similar to one another or members of a family of parts.

   – Objects that might not be in a production configuration (e.g., 'debugging' objects, placeholder objects).

   – Accessibility problems.

   – Assembly problems.

   – Conformance of a physical part to the current design. For example, tools that are in the field must be periodically examined to determine if upgrades are needed.

- Determine status of part metadata to:

   – Show the release status (in-work, released, etc.) of specific parts.

   – Identify parts that have changed recently.

   – Identify parts of a certain type of material, certain weight, etc.

- Observe dynamics in the entire scene (conventionally by animation) to:

   – Understand dynamic interference conditions (e.g., display results from kinematics or mechanisms analysis, vibration, tolerance build-up).

   – Follow flow in systems (e.g., fluid flow in hydraulics tubes).

   – Detect effect of loads, aerodynamic flow, etc., over time.

   – Understand assembly (for manufacturing) and disassembly/reassembly (for maintenance) sequences as shown in Fig. 1.7.

- All of the above assume a single window that displays one style. In addition, it's highly useful to work with multiple versions of the same set of objects to compare the two sets for:

   – Subjective preference (e.g., 'I like the way the hood of that car reflects the lights on the showroom floor.')

   – Net version change.

   – Similarity or difference.

Basically, visual analysis tasks vary widely on what may be an identical picture. In this sense, a technology like massive model visualization is applicable in a number of different situations.

1.2.3   Example Application Domains

In the product manufacturing industry that is increasingly dependent on 3D computer-aided design and virtual prototyping, there are a large number of application scenarios for massive model visualization. Any situation that requires understanding context is appropriate. Although the following examples are derived from the aerospace industry, they are applicable in other industries.

Design reviews. In a design review, a single person or a team representing different viewpoints examines the digital representation for a variety of inconsistencies and to understand the complexity of the overall product.

Part context and location. Figure 1.6 shows one example of a task that's important in a design review, whether an individual or a group performs the review. The task is to find a specific part in a complex environment. The result is understanding the context, and the part could be anywhere in the digital model.

Engineering analysis. As a design matures, a wide variety of engineering analyses are performed that range from computational fluid dynamic to finite element analysis. Visualization provides the key method to understand analysis results in the product context.

Safety. Understanding escape routes and paths gives safety engineers a way to pre-determine the adequacy of a design to accommodate emergency exits.

Survivability. Airplanes are analyzed to determine how well the overall system responds to specific malfunctions.

Massive scans. As designs progress, large scans are conducted to verify that the fabrication process yields parts that conform to the original design. In the automotive industry, such scans (from full-sized clay models) are used as the seed points for the actual digital definition.

Figure 1.6: Find a specific part.

Quality inspection. During assembly and manufacturing, quality inspectors look at the actual product and compare it to the engineering definition. The inspection can occur at any time during the build process.

Assembly instructions. As mechanics assemble a product, the finished product is less important than understanding what it looks like at each step of the assembly process.

Part catalogs. Once delivered, the set of parts that can be ordered is substantial. Freely navigating to specific locations is another way of identifying what replacement parts may be needed and what the surrounding context looks like.

Training. People new to a product benefits from a visual introduction.

Maintenance instructions. In contrast to assembly instructions, maintenance mechanics need to understand the completed visual context to help determine how to take the product apart and what it should look like when re-assembled, as shown in Fig. 1.7.

Tracing systems. In a complex product, systems include wiring, computer networks, hydraulic lines, fuel systems, and ducting. Understanding, assembling, and disassembling the systems often requires tracing from nose to tail. Figure 1.8 provides a systems tracing example. The 3D image is a snapshot, and the actual system complexity is shown if Fig. 1.9.

Sales and marketing. Providing customers a visual image of the entire product, even during its conceptual design phase, aids customer appreciation.

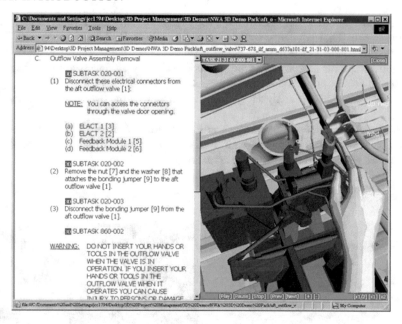

Figure 1.7: Maintenance instruction.

1.2.4   User Performance Expectations

Acceptable interactive performance affects visualization and understanding of massive models more than any other. Performance is not limited to flying through the 3D scene. It also includes model load time and graphical selection with feedback. Users graphically select individual objects as a method of retrieving additional information or as a method of performing other operations on the object.

One of the difficult issues with any application that must display complex, massive models of any type involves model load time. Waiting for something to happen is frustrating, individually or with a group. The faster something appears on the screen, even if incomplete, the more acceptable the solution becomes. For groups, especially when involved in a key review, wait times that last more than a minute can seem like an eternity. The group focuses on the image that isn't there, and idle chatter can last for only so long. Individuals who are working in a multi-window environment tend to be more lenient. As long as other tasks can be accomplished, a wait time of less than 5 minutes is often acceptable. In reality, the closer to 'instant on', the better.

Acceptable 'flying time' (computing new scale, rotate, and translate values in response to some user action and displaying a new 3D image) requires consistent performance. Hand-eye coordination plays a role in determining the acceptable floor for performance. The most commonly accepted value is the flicker fusion threshold. [Wik07] defines the value as 16Hz; [PCM07] suggests 24Hz. In both cases, the value represents the projected display rate for film, video, and other animated methods that cause humans to think that motion is continuous. There is no consideration of eye-hand coordination or that the 'motion' allows the user to better perceive the 3D nature of the model.

The values of 16Hz and 24Hz have been used since projectors first became popular. Acceptable rates vary according to the amount of ambient light. The higher the ambient light, the higher the necessary

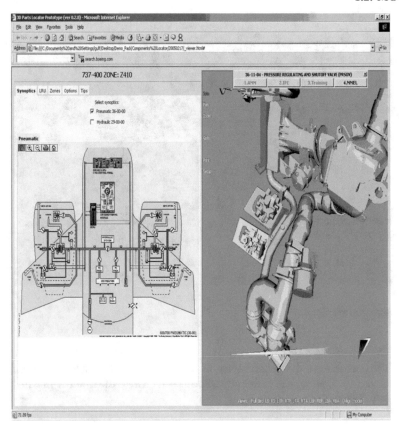

Figure 1.8: Tracing systems.

refresh rate to exceed the flicker fusion threshold. Computer screens have an internal refresh rate of 60Hz or faster to repaint a raster image, whether or not it is changing.

Adding the eye-hand coordination aspect affects the values. Computer game developers work to exceed the screen refresh rate. In undocumented empirical studies conducted at Boeing, massive model visualization users find 16Hz acceptable and 10Hz useful. Any rate under 10Hz diminishes effective interactive flying substantially. When the rate approaches 2Hz, users use the applications grudgingly, if at all.

The final aspect of performance is graphical selection and feedback. This is another case where the impression of instantaneous feedback is essential, and providing feedback for a select action (like a mouse click) requires that feedback appear in less than .25 seconds.

1.2.5   Data Characterization
The evolution of computing has caused an evolution in the level of detail in digital models. Because of performance improvements, users are able to produce data at even faster rates. This is consistent with computing performance in general: users outstrip the capacity of computers at a rate faster than Moore's law improves computer performance.

Figure 1.9: System complexity.

Even with this evolution, the general form of 3D data can be characterized in different ways de-pending on its source. Considering the various data forms can change the manner in which massive model visualization is approached.

1.2.6   Arbitrarily Organized 3D Data

Many industries (e.g., aerospace, architecture, automotive, ship building) have switched from an engineering drawing paradigm to full 3D solid design. The 3D design represents individual parts. Current practice carries the actual design to individual pins on connectors and stencils etched into seats. The complexity has evolved because the 3D data is used as build authority rather than engineering drawings.

While individual parts are spatially coherent, the overall product is not spatially organized. Instead, it represents assemblies and sub-assemblies that are often organized in a seemingly arbitrary manner. For example, the Boeing 777, the first commercial airplane designed entirely using solid modeling techniques, was organized in a manner similar to engineering drawings. Color (as shown in Fig. 1.9) gave engineers an idea about the sort of part in the scene (e.g., blue represented interiors, green structural elements, yellow systems). Each 'model' contained multiple parts that would be placed on an engineering drawing for manufacturing. The 787 has a flat product structure that emphasizes individual parts. Manufacturing information is embedded within the 3D parts, and the level of 3D detail has doubled. Color is used in a more natural, although still not photo-realistic, manner.

3D product data represents numerous systems (e.g., wiring, networks, hydraulic tubes, ducts) that may run from nose to tail. Basically, algorithms cannot assume that individual surfaces and polygons are spatially near one another.

Product data is subject to extensive engineering analysis for stress (finite element analysis), aerody-namics (computation fluid dynamics), and the like. Each form of engineering analysis associates results with the product data. The product data associates different colors whose changing gradient superimposes the engineering results on the product definition or displays the results of dynamic computations [SWS07].

When displaying a scene that shows a huge number of complex objects in a product, there is no guarantee that one object will be physically close to another object. For example, object one may exist in the nose of the airplane, object two in the tail, and object three a tube in the galley. Display algorithms cannot rely on predicting the position of each object unless some additional spatial organization information is provided.

This type of data is similar to data in the animation and games industries. Complicating the situation is that photorealism increases in importance. Photorealism adds model complexity because material characteristics and/or textures be well defined.

1.2.7   Spatially Coherent 3D Data
Significant amount of 3D data are spatially coherent. In other words, 3D polygons or voxels are regularly distributed in space.

Advancements in high-performance and scientific computations generate terabyte data sets that contain simulation data that is spatially coherent. These are high-dimensional data sets (that is, isosurfaces) with hundreds of millions or billions of polygons. For example, Lawrence Livermore Labs generated a 470 million triangle isosurface from a high-resolution 3D simulation of Richtmyer-Meshkov instability and turbulence mixing that has been used extensively as a massive model visualization test case.

Scanning also results in spatially coherent 3D data. The scans can range from tracing the surface anomalies in the Mona Lisa [BGM⁺07] to seismic scans used for oil and gas exploration to CAT scans in medicine to 3D laser scanners used to digitally reconstruct Michangelo's David or a 787 composite fuselage section.

1.2.8   Geographic Coherence
Capabilities like Google Earth and Microsoft Live Earth rely on geographic coherence, a relatively small amount of 3D coordinate data, and a huge amount of texture data. Much of the texture data is obtained from aerial photographs and street level scans. The textures must be correlated to the 3D models that represent the terrain. The terrain itself is geographically coherent on the surface, and the polygon mesh that represents surface topography is usually defined so that the polygons are stored next to one another.

1.2.9   Information Visualization
The basic tenet of information visualization is developing graphical techniques to let a person understand something about the data more clearly. The displays are often algorithmically generated. The algorithms involved are generally tuned to optimize interactive performance.

While the algorithms can help assure spatial coherence, the dynamic nature of generating an information visualization display creates a different set of problems. The previous examples infer a spatial representation by their very nature. Information visualization generally must compute a display during an interactive user session. The difficulty occurs because current massive model approaches often rely on preprocessing the 3D representation to achieve acceptable interactive performance rates.

1.2.10   Implementation Considerations
The enormous size of the 3D datasets poses several implementation challenges. Achieving an implementable delivery system architecture for massive model visualization is a question of balance. [KKF99] provides one example of achieving implementation balance in a system deployed to thousands of workstations. The issue

with massive model visualization must focus on the storage location for the data to make the techniques described in the rest of this book meaningful to more than a handful of users.

In general, the problem is one of copying data from mass storage into CPU RAM or GPU VRAM. The greater the physical distance of the mass storage to the CPU and/or GPU, the slower the transfer becomes. Local disk I/O is slower than USB-attached disk I/O. Server-based storage offers essentially unlimited capacity and depends on a network connection to make the data available for visual processing. Local area networks benefit from fast networks. As a user moves further away from the data, network speed generally decreases and latency always increases. For example, moving tens of gigabytes of data between Puget Sound and Australia becomes a multi-hour task.

Figure 1.10 depicts the system architecture most commonly deployed for massive model visualization. The rendering capability uses only data that is stored on local storage devices. The local storage device is usually a bus-connected, fast hard disk drive. In some cases, it may be a USB-attached hard drive. The local data may be generated on the same user device as is being used for visualization. More often, some sort of synchronization software periodically refreshes the hard drive by pulling new data from servers. Visualization applications like UGSolutions VisMockup, Adobe Reader, Dassault's Digital Mockup Utility, and EnSight Gold assume local data that is displayed with 3D software like OpenGL or Direct3D.

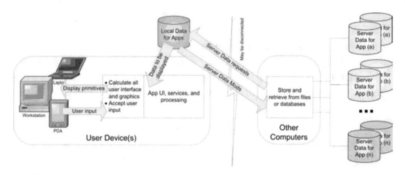

Figure 1.10: Local Data.

The local data architecture works well in most cases. Users can actually work when detached from the network, and performance is reasonably consistent for dynamic interaction. There are two clear disadvantages. First, the synchronization process most often depends on user action to request the latest and greatest data download. Second, a capability to add, modify, or delete data that has been changed since the last refresh is essential. Such a capability avoids transfer of the multi-gigabytes of data across any network, a process that can take hours.

Figure 1.11 introduces the notion of a visualization data cache that is coupled to remote data servers. There are a large number of variations of this architectural form. The most dominant is a Web browser that runs on a user device. The browser automatically checks if data exists in its local cache. If the data is there, the cache automatically checks to see if there is a more recent copy on the server and adds or replaces as appropriate. Other software manages the size of the cache, and data that has not been referenced for a period of time is deleted when the cache would otherwise overflow. Depending on implementation, some of the application processing can actually take place on a fast compute server.

The Remote Data approach avoids the large data download problems in Fig. 1.10. The data servers are generally in the same building or at least on the same campus and interconnected by fast networks. Because multiple compute servers can share the same data servers, configuration management becomes less of a problem. The size of the data cache can vary from zero to being equivalent to the size of the entire data

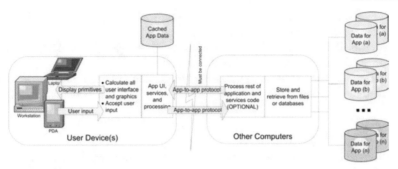

Figure 1.11: Remote Data.

set (as in Fig. 1.10). When the data cache size is zero, display commands (e.g., X-Windows, OpenGL) to be processed on the user device are sent across the network. This approach does not scale well to massive cases because the number of display commands will often exceed the capability of the network to create. Cache capacity becomes a problem when the cache size is less than the size needed for the application data. When this occurs, some of the visualization data can easily be deleted and refreshed when it becomes viewable. The issue becomes the time to download the data from the servers. The cache has to be checked and the data downloaded and processed for inclusion in the frame in 0.1 seconds or less. The net result is that the user experience suffers because of unpredictable performance patterns. If the user has to request that the cache be repopulated manually, the Remote Data architecture transforms into the Local Data architecture.

Figure 1.12 introduces a variant of the Local Data architecture. The idea of a Virtual Terminal (aka Thin Client) moves all heavy processing (including the rendering itself) to a compute server with ties to data servers. The only data sent across the network is user inputs and bit maps. A number of Virtual Terminals are in production use, including Virtual Network Computing, Citrix Metaframe, and HP Remote Graphics.

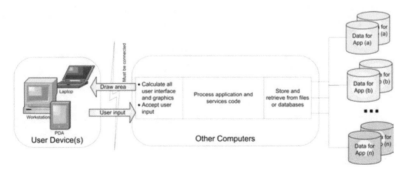

Figure 1.12: Remote Data.

The Virtual Terminal approach scales because bitmap sizes have a physical limit (the window size). Bitmaps can be compressed and analyzed for deltas to decrease network load. Even so, Virtual Terminals must be able to not only compute new frames but also deliver them in 0.1 seconds or less. This means that the time for a new frame must include time for input processing, rendering, preparing and packing the bitmap, physically transmitting it, and then unpacking it. When networks are busy and jitter increases (like transmitting over the general Internet), interframe performance becomes unpredictable and the user experience can suffer for dynamic operations. In extreme cases (e.g., the Australia to Puget Sound round trip), latency itself costs 0.3 seconds, which is 3 times the needed performance of 0.1 seconds or less. Even so, having some capability is often better than nothing at all.

There is no right answer in choosing the correct implementation architecture. Each situation must be assessed on a case-by-case basis.

1.3   CONCLUSION

In many ways, a successful massive model visualization system is no different from other computing systems. As discussed in this chapter, the systems designer must consider the user, the data types, and the system implementation. The differentiator is that the result must achieve real-time performance in spite of the data volume. The rest of this book describes the top three performance bottlenecks and methods to address them. As noted, working these three bottlenecks will reveal a different set of performance issues that must then be addressed. The set of problems involved in massive model visualization will be with the computing and computer graphics communities for the foreseeable future.

CHAPTER 2

Visibility

Visibility algorithms address the problem of determining which surfaces or primitives can be seen from a certain viewpoint or region. In the early 1970s, many of these algorithms were developed to determine the exact hidden portions of the polygons composing a scene in order to generate a rendered image. Later visibility algorithms were extended for shadow determination, global illumination, and interactive display. Determining exact geometric visibility, however, proved to be a very complex problem. Currently, most visibility algorithms employed for massive model rendering focus on computing a quick and conservative approximations of the *visible set*. The visible set contains those primitives that contribute to the current image. The intent is to reject large parts of the scene before the actual per-pixel *visible surface determination* takes place. The aim is to reduce the rendering complexity to the complexity of the visible set of the scene geometry. The process of computing a conservative visible subset of a scene is called *visibility culling* [COCSD03]. Visibility culling and level-of-detail techniques are essential ingredients to create real-time, massive model rendering solutions. Together, these techniques allow the rendering subsystem to have a complexity that depends on image resolution rather than on the source model size.

2.1   INTRODUCTION

Rendering massive models poses important challenges to system developers. This is particularly true for highly interactive 3D applications, such as visual simulations and virtual environments. These applications inherently focus on interactive, low-latency, and real-time processing. Despite the continuing increase in computing and graphics processing power, it is clear to the graphics community that massive datasets cannot be interactively rendered by brute force methods. Therefore, it is important to devise rendering methods that filter out the parts of the dataset that do not effectively contribute to the final image as efficiently as possible.

Visibility culling techniques achieve this goal by detecting which parts can be proved not visible. The three typical visibility culling techniques are *back-face culling*, *view-frustum culling*, and *occlusion culling* (see Fig. 2.1). Back-face and view-frustum culling are local per-primitive (or per-primitive group) operations. These algorithms remove objects whose normal points away from the viewer or whose geometry lies outside of the view frustum. Occlusion culling is generally a more effective technique because it removes primitives that are blocked by groups of other objects. It is more computationally expensive than the first two culling techniques because of its nonlocal nature.

Many different visibility determination strategies have been proposed so far. The approaches are broadly classified into *from-point* and *from-region* visibility algorithms [COCSD03]. From-region algorithms compute a *potentially visible set* (PVS) for cells of a fixed subdivision of the scene. These sets are typically computed offline as part of a preprocessing phase. During rendering, a from-region algorithm renders only the primitives in the PVS of the cell where the observer is currently located. From-point algorithms, on the other hand, are applied online for each particular viewpoint to compute the PVS from scratch and are usually better suited for general scenes, since for general environments accurate PVSs for large viewing regions are hard to compute. From-region methods are mainly used for specialized applications, e.g., urban scenarios or architectural models with large occluders.

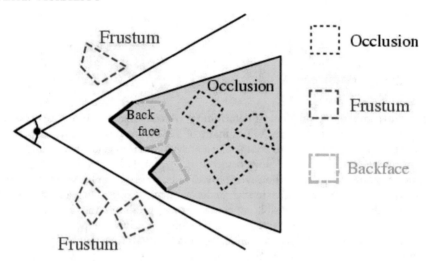

Figure 2.1: **Visibility culling.** Back-face culling, view-frustum culling, and occlusion culling are the most typical examples of methods for computing a visible subset of a scene.

In all cases, some sort of preprocessing is involved. At minimum, preprocessing spatially organizes geometric primitives into a structure that accelerates visibility tests. In the rest of this chapter, we will provide a synthetic overview of the main visibility related techniques for massive model rendering applications.

2.2   OBJECT SPACE SUBDIVISION

Visibility culling methods are typically implemented with the help of a so-called *spatial index*, a spatial data structure that organizes the geometric primitives in 3D space.

2.2.1   Spatial Index Structures

There are two major approaches, *bounding volume hierarchies* (BVHs) and *spatial partitioning*. Bounding volume hierarchies focus on organizing geometric primitives into groups of objects that are encapsulated by a larger and simpler volume. Each volume completely encloses the object groups at lower levels of the hierarchy. The resulting tree can be traversed in a top-down order. Traversal starts at the scene bounding volume defined at the root node. If a bounding volume, i.e., its boundary, is found to be fully or partially visible, rendering continues with its child-volumes. If a volume is completely invisible, traversal of the respective sub-tree can be discontinued because all children will be invisible. Since the focus of BVHs is the organization of primitives, different parts of the hierarchy are not guaranteed to be disjoint.

In contrast to bounding volume hierarchies, spatial partitioning schemes subdivide the scene into a hierarchy of nonoverlapping cells. The scene bounding box is split into disjoint, nonoverlapping partitions. Each partition may further be subdivided in the same fashion. Each atomic partition holds a list of primitives it contains in whole or in part. Processing the partition continues as long as it can be classified as visible.

Quite a number of spatial partitioning schemes have been proposed in the past. The most popular are hierarchical: *hierarchical grids*, *octrees*, and *kd-trees*. More details can be found in [Sam06].

Kd-trees are axis-aligned *binary space partitioning* (BSP) trees. Construction of a kd-tree starts with the bounding box of the model and a list of contained primitives. The scene bounding box is then subdivided into two sub-boxes along one of the three primary coordinate axes. The list of primitives is sorted into the two half boxes and creates two primitive lists, one for each half. Polygons that lie in both halves are either simply replicated or split at the mid plane and distributed into sub-boxes. The process is recursively continued for both sub-boxes and their respective primitive lists. The result is a binary tree in which each node corresponds to a spatial region (called a *voxel*). A node's children correspond to a binary space partition of the parent voxel. If splitting positions are chosen to tightly enclose the scene primitives, kd-trees typically exhibit superior culling efficiency over other acceleration structures.

2.2.2   Generating Spatial Indexes for Massive Models

Even though the concepts of the classic spatial index are simple and well understood, constructing them for massive models requires particular care. Since these models are made of millions of primitives with an uneven distribution and typically do not fit into the main memory, it is important to employ methods that balance construction efficiency with efficacy of the generated structure. On one hand, methods must have a low computational complexity and must use coherent access patterns to avoid I/O thrashing. On the other hand, the methods should still provide optimized space partitions for visibility queries, i.e., they must strive to minimize the expected cost of run-time visibility queries.

2.2.2.1   Kd-trees: the main option of choice for large static models

In case of kd-trees, a de-facto standard for obtaining optimized subdivision is to minimize the cost model for ray-object intersections called *Surface Area Heuristics* (SAH) [Hav00]. This heuristic assumes a uniform distribution of rays with no occlusion. This makes it simple to estimate the probability to traverse the different branches of the hierarchy simply from the surface areas of the bounding boxes of the various nodes. Under this assumption, the expected cost of a particular tree configuration can be estimated as:

$$C_T = K_T \sum_{N \in \text{Nodes}} \frac{SA(V_N)}{SA(V_S)} + K_L \sum_{L \in \text{Leaves}} \frac{SA(V_L)}{SA(V_S)} n_L \,, \tag{2.1}$$

where K_T is the cost of a inner node traversal step, K_L is the cost of a triangle processing step, $SA(V_S)$ is the surface area of the entire tree's bounding box, $SA(V_N)$ is the surface area of the bounding box of inner node N, $SA(V_L)$ is the surface area of the bounding box of leaf node L, and n_L is the number of triangles contained in leaf node L.

An approximately optimal kd-tree is computed by minimizing this cost by performing a top-down greedy optimization. The optimization algorithm recursively splits the model and always chooses the minimum cost split plane at each step. However, a direct implementation of the method is impractical for large models because of the many possible splitting planes required for optimization and the need to sort triangles according along these planes. For these reasons, many authors have proposed simplified techniques for faster tree construction (e.g., [PGSS06, HMS06, SMS$^+$07]). These methods share a common set of concepts. First, they build the hierarchy from axis-aligned bounding boxes of objects instead of individual triangles. Second, they do not test all potential split planes, but only use K heuristically selected, equally spaced planes. In a single streaming pass, triangles are projected into the $K + 1$ "bins" formed by the K planes. Thus, the SAH can be evaluated for the K planes that separate the bins from the triangle counts of the bins.

After computing the SAH for each of these K planes, the best one is selected, and a second linear pass over all triangles subdivides triangles into left and right sub-trees. This approach greatly reduces the number of plane evaluations (K bin planes instead of $O(N)$ triangle bounding planes) and also avoids any sorting. Splitting can thus be done simply with two $O(N)$ passes, and hierarchy is constructed with $O(N \log N)$ operations. Most importantly for massive models, all operations are performed in a streaming fashion with minimal in-core memory demands. The main drawback of these methods is the need to select up front the "right" small set of candidate planes. A more elaborate solution, which avoids binning and considers triangle splitting, is presented in [WH06].

2.2.2.2   BVH: the main option of choice for dynamic models

Bounding volume hierarchies are generally not used for large static environments. They are used for (smaller) dynamic environments for which the hierarchy is either given up-front at modeling time, e.g., by associating bounding volumes to objects in a kinematic hierarchy, or recomputed dynamically as objects move. The research generally focuses on how to update a hierarchy after object motion. Reasonably fast $O(N \log N)$ algorithms for rebuilding BVHs are presented in [WBS07, LYTM06, Wal07]. $O(N)$ methods for refitting an already existing bounding volume hierarchy are presented in [HHS06, WMS06, WK06]. The latter methods are based on the assumption that models undergo small, localized modifications from one frame to the next and that a valid, though not optimal, BVH can be constructed very rapidly by a sequence of small localized modifications of the hierarchy that was valid before the local modification. Also, there is a hybrid method combining both of methods and selectively restructuring portions of BVHs to maximize the performance of rendering [YCM07]. It should be noted, however, that these algorithms, as all dynamic ones in the context of massive models, are still far from being interactive for most complex scenes.

2.3   FROM-POINT ALGORITHMS FOR REAL-TIME VISIBIL-ITY DETERMINATION

From-point algorithms are the basis of all interactive viewing applications. Implementing them in the context of massive model rendering requires special care. In this section, we will first describe the two main visible surface determination approaches employed in massive model rendering applications and then discuss how they can be optimized using local view-frustum and back-face culling techniques or global occlusion culling techniques.

2.3.1   Visible Surface Determination

Visible surface determination aims to precisely determine the surfaces that can a camera can see from a given viewpoint. This is one of the fundamental problems in computer graphics because it is required to produce any synthetic 3D images. It is also known as the *hidden surface removal problem*.

Visible surface determination techniques are essentially methods for solving a sorting problem. The many proposed methods vary in the order in which the sort is performed and how the problem is subdivided to make it more tractable. Currently, only two classes of algorithms are applied when dealing with massive models: rasterization with z-buffering, originally introduced in the early 1970's [Cat74, Str74] and ray tracing [App68], which dates back to the late 1960's. The success of these methods is mainly due to their robustness and conceptual simplicity.

2.3.1.1   Rasterization with z-buffering

Rasterization algorithms combined with the *Z-buffer* are widely used in interactive rendering and are implemented in virtually all modern graphics boards in the form of highly parallel *graphics processing units* (GPUs). Rasterization is an example of an object-order rendering approach (also called forward-mapping): objects to be rendered are sequentially projected onto the image plane, where they are scan-converted into pixels and shaded. Visibility is resolved with the help of the Z-buffer, which stores the distance (or the depth value) of the respective visible object fragment to the observer for each pixel.

This process can be efficiently realized in a pipeline setup, commonly known as the *graphics pipeline*. Early graphics hardware was based on a fixed-function realization of this architecture. Multiple vertex transformation and rasterization units worked in parallel to achieve high throughput. In recent years, graphics hardware has started to feature extensions to the fixed-function pipeline. The generalization allows parts of the vertex transformation and rasterizer stage to be extended. Current GPU pipelines employ a large grid of data-parallel floating-point processors that general enough to implement custom shader functionality. The approach eliminates separate custom processors for vertex shaders, geometry shaders, and pixel shaders. Vertices, triangles, and pixels thus recirculate through the grid rather than flowing through a pipeline with stages of fixed width. Load balancing occurs because the pool of processors can be allocated to each shader type as the graphics load varies.

A rasterization pipeline allows for processing arbitrary numbers of primitives in a stream-like manner. This is especially useful if scenes to be rendered do not fully fit into GPU video memory or in main CPU memory. In its basic form, rasterization techniques work in linear time based on the complexity and number of scene primitives. Linear performance is a direct consequence of the employed object ordering scheme. In order to enable rendering in sub-linear time, spatial index structures must be applied. The structures a-priori limit the number of polygons to be sent down the graphics pipeline. Moreover, since the gap between GPU performance and bandwidth throughout of the memory hierarchy is growing, appropriate techniques must be employed to carefully manage working set size and ensure coherent memory access patterns.

We will see how this works in the next chapters, which present techniques to reduce rendering complexity (Chs. 3 and 4) and managing memory (Ch. 5).

2.3.1.2   Ray tracing

In contrast to rasterization, *ray casting* and its recursive extension *ray tracing* are image order rendering (backward mapping) approaches. Ray tracing models physical light transport with straight lines. A basic ray tracing implementation can be very simple and can be realized with much less effort than a (software) rasterizer. For example, all parts of the rasterizer geometry stage are handled implicitly as a result of the backward projection property. Because of its high computational complexity, ray tracing has been employed in a real-time context only in recent years [WPS+03]. While prototype hardware implementations exist [WSS05], only software ray tracing has so far been applied to massive models.

When it comes to dealing with massive datasets, the underlying issues faced by ray tracing and rasterization approaches are somewhat similar. All the methods have to deal with the same data management and filtering problems and, as we will see, are converging towards proposing similar solutions, based on spatial indexing, data reduction techniques, and data management methods.

2.3.2   View-Frustum and Back-Face Culling

View-frustum and back-face culling are simple but effective from-point operations that can be optimized using spatial data structures for both ray tracing and rasterization. View-frustum culling is usually performed with either a hierarchy of bounding volumes or a spatial data structure such as a kd-tree or an octree [Cla76]. The process can be further accelerated using frame-to-frame coherence as proposed by Slater et al. [SC97] or by simplifying tests for each volume as in [AM00]. Back-facing polygons can be identified with a simple dot product because the polygons' normal points away from the view-point. Hierarchical back-face culling requires additional precomputation of the geometric primitives based on their adjacencies and normal vectors [KMGL99].

For a number of massive models applications, including rendering of dense meshes generated by range scanning, view-frustum and back-face culling are the most effective visibility operations. In this case, scenes have a low depth complexity and, therefore, occlusion culling is often ineffective. For this reason, many dense model visualization systems combine in a single compact data structure all the information used for view frustum culling, back-face culling, level-of-detail selection, and rendering. A common choice is to augment a hierarchy of bounding spheres or axis-aligned bounding boxes with cones of normals [RL00b, CGG+04, GM04]. As we will see, similar hierarchies can also be exploited in systems that include occlusion culling [ISGM02, YSGM04].

2.3.3   Run-Time Occlusion Culling

In order to achieve a sub-linear time complexity on massive models, employing acceleration structures alone and exploiting them for view-frustum and back-face culling is not sufficient for general scenes that contain high depth complexity. It is also necessary to include an early traversal termination in case of occlusion to limit the number of primitives that must be processed by the visible surface determination algorithms. At this point, the spatial indexes discussed in Sec. 2.2 come into play. The spatial indices typically implement hierarchical front-to-back traversal schemes in an efficient manner.

2.3.3.1   Ray tracing

To limit the number of primitives for which the actual ray-primitive intersection test is performed, spatial index structures are necessary. Early traversal termination is simple to implement using hierarchical structures, since visibility is evaluated independently for each ray. Once a hitpoint has been found, it is certain that geometric primitive behind is invisible for that specific ray direction. In interactive ray tracers, such acceleration structures are typically considered to be an integral part of the algorithm and allow for an average logarithmic time complexity with respect to the number of primitives.

An important ingredient that is widely applied in state-of-the-art real-time ray tracing systems is to simultaneously trace bundles of rays called *packets* [WSBW01]. First, working on packets allows for using SIMD vector operations of modern CPUs to perform parallel traversal and intersection of multiple rays.

Second, packets enable deferred shading, i.e., it is not necessary to switch between intersection and shading routines for every single ray. This amortizes memory accesses, function calls, etc. Third, it is possible to avoid traversal steps and intersection calculations based on the bounds of ray packets, which makes better use of both object and scanline coherence. This idea of accelerating ray tracing by working on groups of rays is also exploited in frustum traversal methods [RSH05].

```
void IterativePacketTraverse(ray[4],hit[4]) {
  ( t_near[i], t_far[i] ) = ( Epsilon, ray.t_max );
  // i=0..3 in parallel
  // t_near[i], t_far[i] are the near/far values for the i'th ray
  ( t_near[i], t_far[i] ) = scene.boundingBox.ClipRaySegment(t_near[i], t_far[i]);
  node = rootNode;
  while~(1) {
    while (!node.IsLeaf()) {
    // traverse until next leaf
    d[i]  = (node.split − ray[i].org[node.dim]) / ray[i].dir[node.dim];
    active[i] = (t_near[i] < t_far[i]);
    if for all  i=0..3 (d[i]  <= t_near[i] || !active[i]) {
      // case one, d <= t_near <= t_far for all active rays
      // −> cull front side
      node = BackSideSon(node);
    } else if for all  i=0..3 (d[i]  >= t_far[i]  || !active[i]) {
      // case two, t_near <= t_far <= d for all active rays
      // −> cull back side
      node = FrontSideSon(node);
    } else {
      // case three: traverse both sides in turn
      // correctly update all near/far values
      // push all near/far values for entire packet
      stack.push(BackSideSon(node),max(d[i],t_near[i]),t_far[i]);
      ( node, t_far[i] ) = ( FrontSideSon(node), min(d[i],t_near[i]) );
    }
    }
    // have a leaf now
    IntersectAllTrianglesInLeaf(node);
    if for all  i=0..3 (t_far[i]  <= ray[i].t_closesthit)
      return; // early ray termination
    if (stack is empty)
      return; // noting else to traverse any more...
    // restore all near/far values for entire packet
    ( node, t_near[i], t_far[i] ) = stack.pop();
  }
}
```

To illustrate how this method works, the pseudo-code of BSP packet traversal [WSBW01] is presented in Fig. 2.1. Note that all *x[i]* statements are always executed for all four rays in parallel using a SIMD instruction. Obviously, traversing packets of rays through the acceleration structure generates some overhead. Even if only a single ray requires traversal of a subtree or intersection with a triangle, the operation is always performed on all four rays. Experiments have shown that there is substantial coherence in most scenes, and therefore this optimization pays off very well in practice [WSBW01]. In the future, it is expected that many-core processors will support a high level of data parallelism and SIMD support.

2.3.3.2   Rasterization and occlusion culling

During rasterization, the decision whether traversal of the spatial index can be stopped can also be made in image space by exploiting the Z-buffer. The most recent algorithms exploit graphics hardware for this purpose. During rendering—when the spatial index is traversed hierarchically in a front-to-back order—the bounding box of each visited node is tested against the Z-buffer. Traversal is aborted as soon as occlusion can be guaranteed, i.e., when all Z-values of a box are behind the corresponding stored Z-buffer's values. An efficient implementation of this method requires the availability of fast Z-queries for screen regions. A classic solution is the hierarchical Z-buffer (HZB) [GKM93]. The HZB extends the traditional Z-buffer to a hierarchical Z-pyramid that maintains the farthest Z-value among the corresponding finer level blocks or each coarser block. This allows efficient determination if geometry is during by a top-down traversal of the Z-pyramid.

The hierarchical occlusion map (HOM) method [ZMHH97] is similar in spirit to the HZB. The HOM also supports approximate visibility culling. This is made possible by storing opacity information separately from the distance of the occluders. In this way, the overlap and depth tests can be evaluated independently. To build the HOM, "near" objects are rendered white on black into the frame-buffer. Texturing, lighting, and Z-buffering are turned off before the actual scene rendering takes place. The result is then read back from the frame-buffer, and an opacity pyramid is built bottom-up by performing an averaging operation that is computed using the texture mapping hardware. Testing an object for occlusion in the HOM approach involves first testing whether its projection overlaps some nonblack pixels. A depth test is performed only when the pixel is fully covered

Approximate visibility behavior is controlled by tuning the threshold above which a pixel is considered opaque. This method is more efficient than the original HZB because of the reduced CPU-GPU synchronization needs of the original two-pass approach. A two-pass version of the HZB with a two-graphics-pipeline parallel architecture is implemented in the GigaWalk system [ISGM02]. In this architecture, occluders are rendered on one pipeline and the final interactive rendering of visible primitives takes place on the second pipeline. A separate software thread performs the actual culling using the Z-buffer that results from the occluder rendering. This approach results in a frame of latency in the overall pipeline.

Similar to HOM, the Prioritized-layered projection (PLP) [KS00] implements an approximate culling for the computation of partially correct images in time-critical rendering systems. Preprocessing generates an octree version of the. Each octree cell is assigned a *solidity* value that is proportional to the number of modeling primitives in the cell. During rendering, the algorithm works on budget that attempt to maximize image quality over a fixed amount of polygons or over a specific rendering rate. During traversal, PLP keeps the hierarchy leaf nodes in a priority queue and traverses the nodes from highest to lowest priority. When PLP visits a node, it adds the node to the visible set, removes the node from the queue, and adds the unvisited neighbors of the node to the queue. The priority of a node is computed by initializing it to one. The value is attenuated based on the solidity of the nodes found along the traversal path to the node. This approach mimics the rendering of a semi-transparent volume.

A key feature of the PLP method is that it can estimate the visible set at run time without access to the actual scene geometry. On the other hand, the estimation process does not guarantee image quality, and some frames may show artifacts caused by visible objects not rendered by the method. The method has been improved by augmenting the approximate visible set found by PLP with a conservative method using an item-buffer technique [KS01].

2.3.3.3   Exploiting hardware accelerated occlusion queries

A pure software implementation of the hierarchical Z-buffer is not efficient on current architectures. To some extent, the idea is exploited in the current generation of graphics hardware by applying early Z-tests of fragments in the graphics pipeline (e.g., ATI's Hyper-Z technology or NVIDIA's Z-cull), and providing users with so-called *occlusion queries*.

Occlusion queries define a mechanism by which an application can query the GPU for the number of pixels (or, more precisely, samples) drawn by a primitive or group of primitives. For occlusion culling, the faces of bounding boxes can thus simply be tested for visibility against the current Z-buffer during scene traversal. The occlusion query is used to determine whether or traversal should continue. Although the query itself is processed quickly using the rasterization power of the GPU, the result is not available immediately because of the delay between issuing the query and its actual processing by the graphics pipeline. A naive application of occlusion queries can actually decrease the overall performance because of CPU stalls and GPU starvation. The combined stalls and starvation introduce additional end-to-end latency.

Modern methods exploit spatial and temporal coherence to schedule the issuing of queries [GSYM03, BWPP04, YSGM04, HPB05, KS01] to minimize latency. The central idea of these methods is to issue multiple queries for independent scene parts and to avoid repeated visibility tests of interior nodes by exploiting the coherence of visibility classification.

To illustrate how these method works, the pseudo-code of the coherent hierarchical culling [BWPP04] approach is presented in Fig. 2.2. The basic idea behind the method is to avoid testing for occluded nodes that passed the occlusion culling test in the previous frame. The algorithm visits the hierarchy in a front-to-back order and immediately traverses any previously visible interior node in a recursive manner. For all other nodes, the algorithm issues an occlusion query and stores it in a queue. If the node was a previously visible leaf node, it also renders the primitives in that node immediately without waiting for the query result. As soon as the query result becomes available, the result is read, and the algorithm stops at the node if it is fully occluded. Otherwise, traversal continues recursively. In either case, the visibility status of queried nodes is pulled from the hierarchy. A node is marked as visible as soon as one of its children is not totally occluded. The method is simple and works well, even in fully dynamic scenes. It has been later improved by [GM05] and [CBWR07] to integrate level-of-detail selection (see Ch. 4).

2.4   FROM-REGION ALGORITHMS FOR PREPROCESSED VISIBILITY DETERMINATION

Visibility preprocessing is an important method to accelerate real-time walkthroughs of large scale virtual environments. Traditional visibility preprocessing algorithms assume that, in addition to partitioning the object space into a set of objects, the view space is partitioned into a set of view cells. During preprocessing, the algorithms determine a potentially visible set of objects (PVS) for each view cell. At run-time, only the PVS stored with the view cell containing the viewpoint needs to be considered for rendering. This can lead to large savings in rendering time.

While exact visibility from a single viewpoint can be calculated using visible surface determination methods, computing the PVS for a region is much harder. Excellent algorithms for computing exact visibility from a region in space exist for general scenes [Dur99, DD02, NBG02, Bit02, HMN05, MAM05]. However, their running time and memory costs make them very hard to apply to massive models. Deciding whether an object O is visible from a region R requires detecting whether there exists at least a single ray that leaves R and intersects O before it intersects an occluder. Since there are four degrees of freedom in the description of a ray in three-space, the problem is inherently four-dimensional [Tel92, DDTP00].

For this reason, many authors have concentrated on "conservative" techniques, i.e., techniques that simplify computation by (hopefully) slightly over-estimating the PVS. Over-estimation includes some objects that are actually invisible and never excludes completely unoccluded objects. In reality, this problem is

```
TraversalStack.Push(hierarchy.Root);
while (not TraversalStack.Empty() or not QueryQueue.Empty()) {
  //-- PART 1: process finished occlusion queries
  while (not QueryQueue.Empty() and (ResultAvailable(QueryQueue.Front()) or TraversalStack.Empty()) ) {
    node = QueryQueue.Dequeue();
    if (GetOcclusionQueryResult(node) > VisibilityThreshold) {
      if (IsLeaf(node)) {
        Render(node);
      } else {
        TraversalStack.PushChildren(node);
      }
      while (not node.visible) {
        node.visible = true;
        node = node.parent;
      }
    }
  }
  //-- PART 2: hierarchical traversal
  if (not TraversalStack.Empty()) {
    node = TraversalStack.Pop();
    if (InsideViewFrustum(node)) {
      wasVisible = node.visible and (node.lastVisited == frameID - 1);
      leafOrWasInvisible = not wasVisible or IsLeaf(node);
      node.visible = false;
      node.lastVisited = frameID;
      if (leafOrWasInvisible) {
        IssueOcclusionQuery(node);
        QueryQueue.Enqueue(node);
      }
      if (wasVisible) {
        if (IsLeaf(node)) {
          Render(node);
        } else {
          TraversalStack.PushChildren(node);
        }
      }
    }
  }
}
```

also very hard, and there are practically no published, provably conservative techniques for general environments. Instead, published techniques describe techniques restricted to particular types of scenes. Examples include the limitations for architectural building interiors [ARB90, TS91], 2.5D visibility for terrains and urban scenes [WWS00, BWW01, KCCO01], volumetric occluders [SJDS00], or large occluders close to the view cell [DDTP00, ASVN00, LSCO03]. For general scenes, nonconservative sampling based solutions that compute from-region solutions that combine results from from-point queries have recently emerged as a practical approach because of their robustness and ease of implementation.

2.4.1   Specialized Conservative Solutions

The first visibility preprocessing methods designed for indoor architectural environments [ARB90, TS91, LG95] partition the scene into cells roughly corresponding to rooms in a building (see Fig. 2.2). The cells are connected by portals which correspond to transparent boundaries between the cells and roughly correspond to windows or doors. They first subdivide space into cells using a 2D BSP tree where walls become split planes. The next step saves the collection of other potentially visible cells in each cell. Visible cells become those which can be seen through the set of portals. These early methods are very conservative but still effective for architectural walkthrough applications. They have been shown to work quite well for architectural buildings. In addition to being used for view culling, the structuring of the database into cells can be used to optimize data access, since it is easy to pre-fetch soon-to-be-visible parts of the scene.

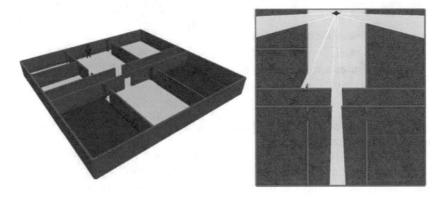

Figure 2.2: **Cells and portals.** Cells only see other cells through portals. All portals in the viewing cell are tested against the view frustum planes and accepted, discarded, or clipped as necessary. Only objects in the areas highlighted in yellow need to be rendered.

These methods work well in practice with simple building layouts based on orthogonal walls. In scenes with nonaxial polygons, the subdivision into cells and portals may result in scene fragmentation [TS91]. Even though a number of automatic solutions for dealing with this problem have been presented [LCCO06], manual construction of cells and portals structures in the modeling phase is still very popular, especially in the video game industry.

There are other methods related to cells-and-portals techniques. The techniques rely on the presence of large occluders in the scene, a characteristic of many indoor scenes. Coorg and Teller [CT96, CT97] and Hudson et al. [HMC+97] presented early approaches. The approaches are based on exploring visibility relationships between two convex objects to determine regions of space where one occludes the other. In their methods, a certain number of occluders are used to prune portions of the scene hierarchy. During preprocessing, the scene is partitioned into view cells, which store the occluders that will be used whenever the viewpoint is inside the cell. Occlusion computation is performed using the shadow frustum of the selected occluders (see Fig. 2.3).

Bittner et al. [BHS98] improved the above occluder-shadow algorithms by combining the shadow frusta of the occluders into an *occlusion BSP tree*. This way redundant (already occluded) occluders may be removed and each bounding volume in the spatial hierarchy is tested against the tree instead of testing each one of the shadow frusta. These methods and similar methods that focus on single large occluders [COFHZ98, COZ98, WBP98] are essentially from-point methods even if some explicit cell information is stored. Lack of occluder fusion is a serious limit because single objects can become effective occluders in the from-region case only if the size of the cell is very small. This result is demonstrated in [NFLYCO99]. For this reason, much of the later work has focused on occluder fusion, a technique that merges many small

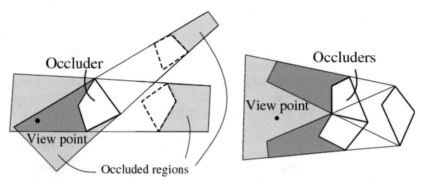

Figure 2.3: **Occluder shadows.** Occluded areas and simple occluder fusion in methods based on large occluders.

occluders for visibility computation [ASVN00]. In general, however, *occluder fusion* proves difficult in the context of provably conservative algorithms.

Another situation in which specialized conservative solutions can be applied are outdoor urban scenes and digital elevation models. Both classes of models are typically considered 2.5D by nature. In these situations, from-region visibility can be computed by appropriately composing simpler 2D visibility queries. Stewart [Ste97] proposed an early conservative hierarchical visibility algorithm that precomputes occluded regions for cells of a digital elevation map. Wonka et al. [WWS00] observed that it is possible to compute a conservative approximation of the visibility of an object in 2.5D from discrete point samples placed on the boundary of a view cell. The test determines if the occluders are shrunk by an amount corresponding to sample spacing (see Fig. 2.4). This method is limited to volumetric occluders.

Koltun et al. [KCCO01] transforms the 2.5D problem to a series of 2D visibility problems solved using dual ray space and the Z-buffer algorithm. Bittner et al. [BWW01] use a line space subdivision maintained by a BSP tree to calculate the PVS.

2.4.2 Aggressive Occlusion Culling Using Visibility Sampling

Given the inherent complexity of visibility computations, today's tools for PVS computation in general scenes are almost universally based on sampling. These tools typically do not guarantee that all visible objects are included in the computed PVS.

The typical solutions use different approaches. Some sample visibility by randomly selecting a number of rays covering the view space and stopping when the probability of missing a visible object is considered low enough [ARB90, SGwHS98, SJDS00]. Others sample the boundary of the view cell first and then sample visibility from each of these points [LH96, Stu99, GM05]. Yet another approach shoots rays from the scene triangles towards the view cell [GSF99]. This last option leads to ray space oversampling for most scenes that contain densely occluded scenes.

One of the major advantages of the sampling approach to visibility is that it can harness the advances in visible surface determination using from-point algorithms, which leads to fast and robust solutions.

Nirenstein and Blake [NB04] proposed an approach which uses rasterization hardware for sampling visibility. The method generates a set of random sample points inside the view cell and rasterizes the scene into the faces of cubes centered at sample points. Each of the six sides of these cubes are treated as independent depth and frame buffers onto which the scene is rendered. Each polygon is then assigned a distinct 32 bit color. The set of polygons mapped by at least one pixel in any of the six frame buffers is

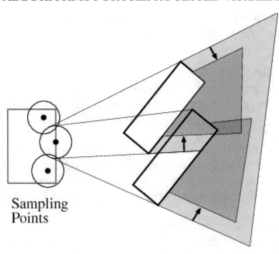

Sampling
Points

Figure 2.4: **Conservative visibility by sampling in 2.5D.** A conservative approximation of the visibility of an object from discrete point samples placed on the boundary of a view cell can be obtained if the occluders are shrunk by an amount corresponding to sample spacing (adapted from [WWS00]).

considered to be the set of polygons visible from the sample point. The visibility cube can be considered a high density sampling over the angular domain for a fixed spatial position. By combining the visibility from a number of these cubes, from-region visibility can be estimated. Sample points are generated adaptively using a subdivision heuristic based on visibility sample similarity. This is derived from the simple observation that if two viewpoints see similar item-buffer images, then any viewpoint between the two will also most likely see a similar image.

A different approach that focuses on harnessing a fast ray tracing kernel has been recently presented by Wonka et al. [WWZ+06]. The proposed solution is based on stochastic ray shooting and takes as input triangular models without connectivity information. The two main components of the method are: (1) a sample generator for exploring the ray space with independent random samples and a sampling queue for propagating the ray using adaptive border sampling and (2) reverse sampling strategies to mutate a ray's origin or direction in response to previously encountered hit points. Each time a ray hits a triangle not previously encountered, it adds it to the PVS and tries to find neighboring ones by mutating the ray's direction toward predicted positions on the exterior side of the triangle's edges. This adaptive border sampling explores connected visible primitives of the input model from a single viewpoint but cannot penetrate into gaps visible only from other portions of ray space. This situation is handled by the reverse sampling strategy. If the actual hit point of one of these mutated rays is much closer than the predicted one, a new blocker is identified. A mutated ray from a different view cell position to the predicted hitpoint is then created so that it passes by the occluding triangle.

These sampling based strategies have many similarities but also have characteristics that make them appropriate for different applications. The rasterization-based solutions tend to consistently underestimate the visible set because of the discretization performed by the cube sampling step and the highly different spatial resolution and angular resolution. The major target is thus for previewing applications that can tolerate minor image-space errors in exchange for fast preprocessing and rendering times. By contrast, the ray tracing solution tries to obtain the best possible approximation of the visible set and is thus more appropriate for a wider range of applications.

2.4.3   View Space Subdivision Strategies

Even though all from-region algorithms rely on a subdivision of view space into cells, the problem of how to subdivide view space into view cells has received only marginal attention to date [MBW06]. In particular, most methods assume that view space is subdivided, automatically or by user intervention, before PVS computation takes place.

Classic cells and portals solution rely either on manual construction or a priori knowledge of the scene to construct the cell and portal graphs [LCCO06]. Most algorithms targeting general scenes also assume that the view cells are either defined by the user or use simple view space subdivisions. The most popular subdivision approach is the regular grid. Only a few methods use visibility computations for guiding view space subdivision.

Gotsman et al. [GSF99] constructs a top down kd-tree subdivision of the view space that uses sampled visibility to evaluate the efficiency of the candidate splitting planes. Van de Panne and Stewart [vdPS99] proposed an alternative approach that merges view cells bottom up if the associated PVSs are similar. Recently, Mattausch et al. [MBW06] proposed a method that aims to minimize the estimated rendering cost for a given view space partition. The method has been later improved to combine viewspace and object space subdivision into a single system [MBWW07]. These adaptive methods are able to produce compact and effective view space partitions automatically. However, because of their memory and computational costs they have been so far applied only to small or medium-size models and have not been applied to massive models with hundreds of millions of polygons

An alternative approach that exploits a particular form of from-region visibility is provided by the preprocessing component of the Far Voxels [GM05] massive model rendering framework. Far Voxels uses model-space partitioning to define a particular hierarchical view-space partitioning. The system partitions the model into cells using a BSP constructed according to the surface-area heuristic and constructs for each inner node a discretized level-of-detail representation consisting in a regular voxel grid (see Ch. 4). In order to guarantee that voxels always subtend a very small viewing angle, the level-of-detail strategy guarantees at run-time that a node is only going to be displayed from viewpoints farther than a given distance d_{min}. This fact is exploited in the preprocessing by a from-region occlusion culling strategy. In this case, the view-space region is not a finite cell, but rather the dual of the cell delimited by the surface S at distance d_{min} from the node's bounding volume (see Fig. 2.5). As reported in [GM05], this visibility preprocessing strategy is extremely effective for complex models, as environmental occlusion leads to eliminate a large portion of the voxels (over 40% for the Boeing 777 dataset). More aggressive culling is then performed at run-time by further application of a from-point strategy based on hardware occlusion queries.

2.4.4   Dealing with the PVS Storage Problem

There is an obvious trade-off between the quality of the PVS estimation and memory consumption and precomputation time. Smaller view cells not only improve the quality of PVS computation but also increase the number of view cells that need to be precomputed. In addition to requiring large precomputation times, a large number of view cells can result in extremely large storage requirements for storing all PVSs. The large size increases the bandwidth required to communicate the PVSs to the rendering engine.

The storage and bandwidth problem can be tackled by using compression techniques for storing and transmitting the PVS [vdPS99, GSF99, COFHZ98, COZ98, NFLYCO99] or by using intermediate representations from which PVSs can be rapidly reconstructed at run-time [KCCO00]. Another solution, which also reduces precomputation costs, is to compute the PVS of regions dynamically during the walk-

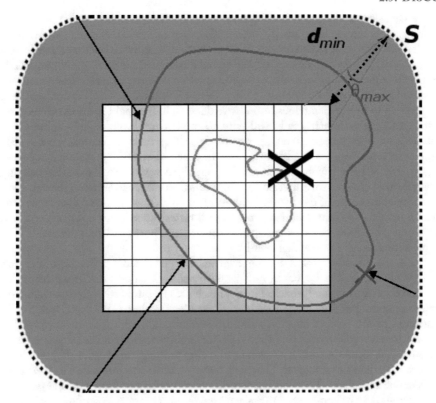

Figure 2.5: Precomputed visibility in the Far Voxels framework. Environmental occlusion is taken into account to remove always occluded voxels and to restrict the sampling to potentially visible surfaces. In the image, the blue object hides the yellow one, and only gray voxels are considered unoccluded (adapted from [GM05]).

through [KCCO01, WWS01b]. The basic idea behind these techniques is to asynchronously compute a PVS for a single cell around the viewer in a parallel thread that runs locally or on a separate server.

This approach essentially combines the advantages of run-time from-point visibility processing and preprocessed from-region visibility calculations. The method works well if the viewer's position is easily predictable and the maximum viewer movement speed is known in advance. It should be noted that, while asynchronous solutions have also been presented for purely from-point algorithms [ISGM02, GSYM03], the visibility results are only valid for one frame and most appropriate to a pipelined architectures that couples visibility and rendering frame rates. Finally, the approach introduces one frame of latency in the overall pipeline.

2.5 DISCUSSION

In this chapter we have reviewed a number of approaches to visibility suitable for massive model rendering applications.

We have seen that, at the broad level, current visibility determination approaches can be classified into rasterization or ray tracing methods. The main advantage of rasterization algorithms is the ability to efficiently exploit scanline coherence. Consequently, such techniques work best in cases where large screen space areas are covered by a few triangles. Conversely, ray tracing and ray casting can perform theoretically and asymptotically better if visibility needs to be evaluated point-wise. In real implementations, however, the main overhead comes from the memory access pattern of ray tracing, which is nonlocal. In addition, it is hard or impossible to fit the hierarchies for massive datasets in the main memory. Out-of-core accesses during ray traversals are extremely costly and hard to implement. Even though some out-of-core ray tracing solutions have been presented, e.g., [WDS04, YLM06], their realization is more elaborate than corresponding rasterization approaches because of the finer granularity of external memory access requests. Interestingly, hierarchical front-to-back rasterization combined with occlusion culling techniques can be interpreted as a form of beam or frustum tracing. This blurs the difference between the two approaches. As of today, most systems use one of the two techniques exclusively. It is, however, likely that future systems may incorporate hybrid combinations of ray tracing and rasterization. This approach will become more prevalent as graphics hardware becomes more and more general purpose. More general GPUs will allow for executing rasterization and ray tracing side-by-side.

This trend has already started in the context of specialized applications. For instance, the BlockMap [CDBG⁺07] system for urban models decomposes a city model into a hierarchy of blocks that are then rasterized from front-to-back. The approach uses hierarchical occlusion culling method that exploits occlusion queries but then renders each primitive block using a GPU ray-caster.

Independent of the type of selected visibility determination approach, massive model rendering applications have to select the most appropriate visibility culling method for the particular model domain. In general, from-point techniques are more robust and easy to integrate because they require less storage and less preprocessing time and resources. There are, however, situations in which a from-region algorithm provides considerable advantages. From-region algorithms are used for a number of video games. Scenes are modeled only once and can be constructed to make region selection easy. A good from-region algorithm for general models with reasonable preprocessing cost and good storage optimization remains an open issue.

It is also worth noting that visibility culling methods have many applications other than accelerating rendering. In particular, they are at the core of shadow computations and global illumination.

2.6   FURTHER READING

Visibility computations are fundamental techniques in a number of areas. Surveys that overlap with our work exist. Samet has written an exceptionally thorough coverage of spatial data structures and applications [Sam06]. Cohen-Or et al. [COCSD03] and Bittner and Wonka [BW03] provide surveys of visibility in computer graphics with a focus on culling approaches for 3D real-time rendering.

Simplification and Levels of Detail

Relying on efficient visibility determination alone is not sufficient to ensure interactive performance for highly complex scenes with a lot of very fine details. In such cases, many visible modeling primitives may only project to a single pixel or sub-pixel. In order to bound the amount of data required for a given frame, a filtered representation of details must thus be available. Computing such a representation from highly detailed models and efficiently extracting the required detail from this representation at rendering time is the goal of *complexity reduction techniques*, which are discussed in this chapter.

3.1 INTRODUCTION

One of the major problems in massive model rendering is how to obtain representations that satisfy both accuracy and timing constraints. Unfortunately, in the general case, the more accurately a digital model represents a real object, the more complex becomes its representation and consequently the higher its rendering cost. This is particularly true when the boundary of an object is represented by a piecewise polygonal surface, in most cases a triangle mesh, or a set of point samples. As a general rule, one can assume that with a curved object surface, the more refined is the discretized model, the higher its accuracy.

One straightforward approach to meet performance dirstraints is to simplify complex models until they become manageable by the application: if models are too complex, make them simpler! This static "throw-away-input-data approach" might seem too simplistic, but can be considered beneficial in a number of practical use cases. A common application of static simplification is reducing the complexity of very densely over-sampled models. For instance, models generated by scanning devices and iso-surfaces extracted by algorithms such as marching cubes are often uniformly over-tessellated because of the nature of most reconstruction algorithms. By adaptively simplifying meshes so that local triangle density adapts to local curvature, it is often possible to radically reduce triangles without noticeable error. More generally, users may want to produce an approximation which is tailored for a specific use, e.g., viewing from a distance.

In the more general case, however, the quality loss incurred when using off-line simplification techniques is not acceptable, and the application must resort to more general adaptive techniques able to filter data at run-time. These level-of-detail (LOD) techniques reduce memory access and rendering complexity by exploiting multi-resolution data structures for dynamically adapting the resolution of the dataset (the number of required model samples per pixel). They complement the visibility culling techniques reviewed in the previous chapters and reach the same goal by avoiding processing parts that can be proved not visible because out of view (in the case of view-frustum culling) or masked (in the case of back-face and occlusion culling).

In this chapter, we will briefly review techniques for reducing model complexity. We will first focus on how to effectively simplify massive triangle meshes, which are by far the most common representation of geometric models. We will then discuss how simplified representations can be arranged in a multiresolution structure to give the visualization application the ability to determine the amount of detail required at different parts of the model at run-time. In the next chapter, we will discuss how the complexity of the rendering operation can be also reduced by switching to representations other than triangle meshes.

3.2   GEOMETRIC SIMPLIFICATION

Geometric simplification is a well-studied subject, and a number of high-quality automatic simplification techniques have been developed [Lue01]. Optimal approximation of a surface, in terms of computing the minimal number of triangle primitives that would satisfy some approximation error metric, is known to be NP-Hard [AS94]. Hence, most research has focused on developing heuristic methods. A large body of research exists in this area. It is important to note, however, that early mesh simplification efforts have focused on in-core meshes. It is only from the early 2000s that effective techniques have started to appear for high-quality adaptive simplification of polygonal meshes that are too large to fit in-core.

3.2.1   Global and Local Mesh Simplification Strategies

At the broadest level, simplification methods may be grouped into *global strategies* that are applied to the input mesh as a whole, and *local strategies* that iteratively simplify the mesh by the repeated application of some local operator.

Two prominent examples of global strategies are *spatial vertex clustering* [RB93] and *variational shape approximation* [CSAD04]. Spatial clustering approaches are based on the concept of spatially partitioning the input vertex set into clusters, unifying all vertices within a given cluster, and removing all degenerate faces and duplicate vertices in the process. These methods are, in general, fast and appropriate for very large meshes, since they can be easily implemented in an out-of-core fashion [Lin00]. On the other hand, vertex clustering makes it difficult to produce meshes with a prescribed number of primitives, can drastically alter the input mesh topology in an unpredictable manner, and may not produce very faithful geometric approximations at low levels of detail. The other hand of the spectrum is the shape approximation approach [CSAD04], which casts mesh simplification as a global optimization problem. The basic idea is to employ a variational partitioning scheme to segment the input mesh into a set of nonoverlapping connected regions, then fit a locally approximating plane to each region, and finally re-mesh it. The obtained approximation is of high quality. The quality comes at the cost of an expensive optimization process that is difficult to implement for very large meshes.

Local strategies are by far the most common simplification approaches because of their efficiency and robustness. The wide majority of the simplification methods iteratively simplify an input mesh by sequences of vertex removal or edge contraction operations (see Fig. 3.1). In the first case, originally introduced by Schroeder [SZL92], a vertex is removed from the mesh at each simplification step and the resulting hole triangulated. In the second case, popularized by Hoppe [Hop96], the two endpoints of an edge are contracted to a single point and the triangles that degenerated in the process are removed. Edge collapse is the most popular atomic simplification action because it does not require explicit triangulation of the area affected by the action and because applications can place the new vertex in an optimal position. One nice property of local iterative algorithms is that they allow the user to easily specify termination criteria for the simplification. For example, the user may allow the algorithm to run until the mesh contains k faces or until the error at a given vertex exceeds some threshold. Global strategies, in contrast, are less amenable to this level of specific control. Even if there are some global strategies with similar properties, e.g., simplification envelopes [CVM+96], it is hard to scale them to massive models.

3.2.2   Controlling Approximation Accuracy

The control of the *approximation accuracy* is critical in the process, if only to guide the order in which to perform operations.

The error measures most frequently used in the literature are based on the L_∞ norm [CVM+96, KLS96, CCMS97] or the L_2 norm [Hop96, HDD+93]. The error evaluation method most frequently used

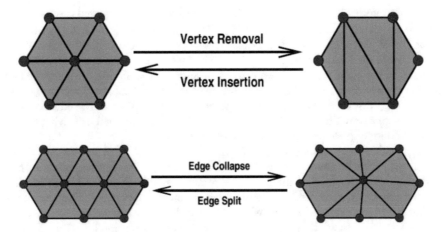

Figure 3.1: Mesh simplification operations and their inverses. Top: a vertex is removed and the resulting hole triangulated. Bottom: an edge is collapsed to a single point.

in current applications is the *quadric error metrics* (QEM), originally proposed by Garland and Heckbert in [GH97].

The approach is an efficient technique for approximating the *vertex_to_plane* distance, that, for each vertex v of a simplified mesh, computing the distance from the set of planes corresponding to the section of the input mesh M in the proximity of v. Instead of explicitly storing the plane sets, as in [RR96], QEM represents the error by a *quadric matrix*. At the initialization stage, a quadric matrix is assigned to each vertex v. This matrix represents the sum of squared distances to the planes of the faces incident at v. The error of v is therefore:

$$\varepsilon_v = \sum_{p \in \text{planes}(v)} \left(p^\top v \right)^2 \tag{3.1}$$

where $p = [\, a\ b\ c\ d\,]^\top$ represents the plane defined by the equation $ax + by + cz + d = 0$ where $a^2 + b^2 + c^2 = 1$. This error metric can be rewritten in quadratic form as follows:

$$\varepsilon_v = \sum_{p \in \text{planes}(v)} \left(v^\top p \right)\left(p^\top v \right) \tag{3.2}$$

$$= \sum_{p \in \text{planes}(v)} v^\top \left(pp^\top \right) v \tag{3.3}$$

$$= v^\top \left(\sum_{p \in \text{planes}(v, M_i)} K_p \right) v, \tag{3.4}$$

where K_p, called *fundamental error quadric*, is the matrix:

$$K_p = pp^\top = \begin{bmatrix} a^2 & ab & ac & ad \\ ab & b^2 & bc & bd \\ ac & bc & c^2 & cd \\ ad & bd & cd & d^2 \end{bmatrix}. \tag{3.5}$$

The important points to note are that K_p can be used to find the squared distance of any point in R^3 to the plane p and that the distance from a set of planes P can be simply obtained by summing together the quadrics of all the planes in P. Thus, the quadric associated with any initial vertex is the sum of all the quadrics relative to the incident planes. Similarly, when we evaluate the error associated with an edge collapse, we first join the plane set of the two vertices v_1 and v_2 by summing the associated quadrics and then evaluate the squared distance by multiplying the quadric with the collapsed vertex v_{new}. The position of the new vertex v_{new} can be either one of the vertices, the middle point of the edge, or a new location that minimizes the approximation error [GH97]. The latter can be computed by solving a linear problem, since the error function to be minimized is a quadratic function.

Summing quadrics may introduce some imprecision if the set of planes are not disjoint because replicated planes will then be counted several times. On the other hand, using quadrics has the major benefit of reducing space overhead. Only a small symmetric matrix is stored per vertex. This leads to an extremely fast error metric consisting of simple vector–matrix operations. The method is thus very effective and has later been extended to deal with vertex attributes such as color, texture coordinates, and normals [GH98, Hop99b]. In addition, the method works in any dimension [GZ05].

3.2.3   Simplifying Massive Meshes

Various techniques have been presented in recent years to face the problem of massive mesh simplification. With the exception of memoryless clustering approaches [Lin00, Lin03, SG01], which are based on a global simplification strategy approach providing little control on simplification quality, in most current systems, simplification is performed in an iterative greedy fashion, which maintains a sorted list of candidate operations and applies at each step the operation associated to the minimal simplification error. Unfortunately, a direct implementation of this approach is not well suited to work on massive meshes, since maintaining a priority queue of possible operations results in a memory consumption proportional to the size of the original mesh, a clearly untenable situation for extremely large models: even if this obstacle could be overcame by using out-of-core data structures, the order of contraction operations could exhibit little locality in terms of memory accesses, with detrimental effects on algorithm performance. The two main solutions that have been proposed for these problem are *streaming simplification* methods or *mesh partitioning* methods.

3.2.3.1   Streaming simplification approaches

The key insight behind streaming simplification [WK03, ILGS03b] is to keep input and output data in streams that interleave connectivity information with vertex, edge, and triangle properties. Finalization information marks when all triangles around a vertex or all points in a particular spatial region have already been defined in the stream. This representation allows for streaming very large meshes through main memory while maintaining information about the visiting status of edges and vertices. Only a small portion of the mesh is kept in-core at any time while the bulk of the mesh data resides on disk. Mesh access is restricted to a fixed traversal order, but full connectivity and geometry information is available for the active elements during traversal. For simplification, an in-core buffer is filled and simplified, and output is generated as soon as enough data is available.

3.2.3.2   Mesh partitioning approaches

In contrast, mesh partitioning methods are based on iterative simplification of mesh regions. Several authors [Hop98, Pri00] have proposed methods in which a mesh is segmented so that each piece fits in main memory. The pieces are then simplified in-core. The boundary edges are preserved so the segments can be rejoined. This process is iterated and new boundary edges chosen for each iteration. While this solution is conceptually appealing, the segmenting and rejoining operations are expensive and make this approach less attractive for very large meshes. In particular, some boundaries remain unchanged until the very last simplification step, potentially decreasing simplification quality and leading to performance problems for very large meshes.

OEMM [CMRS03] avoids the region boundary problem by exploiting an out-of-core data structure that maintains relationships between blocks and thus supports simplification of block boundaries. The structure used for mesh partitioning is an out-of-core octree. For simplification, instead of keeping a global heap with all the possible collapses, the OEMM is traversed following the lexical order of the leaves. For each subtree that is loaded into memory, a local priority queue is created. The mesh is simplified separately from the rest and eventual changes are propagated to neighboring cells. At the end of the traversal the mesh is uniformly simplified.

Another efficient technique for avoiding boundary locking has been proposed by [CGG+04, CGG+05]. In these approaches, the mesh is spatially partitioned by hierarchical volumetric subdivision schemes that create conforming volumetric meshes to support local refinement and coarsening operations. These methods also have the capability of producing continuous LOD representations and are discussed in Sec. 3.3.4.

3.3   LEVEL-OF-DETAIL

A *level-of-detail* (LOD) model is a compact description of multiple representations of a single shape and is the key element for providing the necessary degrees of freedom to achieve run-time adaptivity. LOD models can be classified as *discrete*, *progressive*, or *continuous* LOD models.

3.3.1   Discrete LOD Models

Discrete LOD models simply consist of ordered sequences of representations of a shape, where each member represents an entity at increasing resolution and accuracy. The expressive power of discrete LODs is limited to the different models contained in the sequence: there is usually a small number in the sequence to save space and their accuracy/resolution is predefined (in general, it is uniform in space). Thus, the possibility of adapting dynamically to the needs of user applications is small. The extraction of a mesh at a given accuracy is reduced to selecting the mesh whose characteristics are the closest to the application needs. Such models are standard technology in graphics languages and packages, such as VRML or X3D and are used to improve efficiency of rendering: depending on the distance from the observer or a similar measure, one of the available models is selected. The approach works well for small or distant isolated objects, which can be found in some kinds of CAD models [EMB01]. However, it is not efficient for large objects spanning a range of different distances from the observer. Since there is no relation among the different LODs, there are no constraints on how the various detail models are constructed.

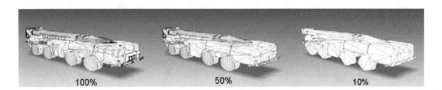

Figure 3.2: Discrete LODs. Three levels of detail of an object constructed by polygon decimation.

3.3.2   Progressive LOD Models

Progressive models contain a coarse shape representation and a sequence of small inverse decimation operations (e.g., edge splits). When the edge splits are applied to the original coarse representation, they produce finer representations at intermediate levels of detail. Such models lead to compact data structures because all modifications in the sequence belong to a predefined type, and thus can be described with a few parameters. The most notable example is the Progressive Mesh representation [Hop96], which has been part of the DirectX library since version 8. In this case, the coarsening/refinement operations are edge collapses and edge splits. A mesh at uniform accuracy can be extracted by starting from the initial one, scanning the list of modifications, and performing modifications along the sequence until the desired accuracy is obtained. As in the case of discrete LODs, the approach works well for small or distant isolated objects, but it is not efficient for large objects spanning a range of different distances from the observer.

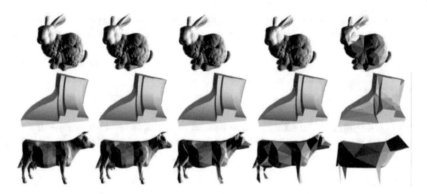

Figure 3.3:   Progressive LODs. Sequence of approximations of the bunny, cow, and fandisk datasets extracted from a progressive mesh. The original models on the left have full resolution. The approximations to the right have 20%, 10%, 5%, and 1% triangles, respectively. Image courtesy of CRS4.

3.3.3   Continuous LOD Models

Continuous LOD models improve over progressive models by providing full support for selective refinement, i.e., the extraction of representations with an LOD that can vary in different parts of the representation. This allows new models to be visualized on a virtually continuous scale. Continuous LODs are typically created using a refinement/coarsening process similar to the one employed in progressive meshes. However, rather than just storing a totally ordered sequence of local modifications, continuous LOD models link each local modification to the set of modifications that defined it. Thus, in contrast to progressive models, local updates can be performed without complicated procedures to find out dependency between modifications. A general framework for managing continuous LOD models is multi-triangulation [FMP98]. Multi-triangulation is based on encoding the partial order that describes mutual dependencies between updates as a directed acyclic graph (DAG). Each node in the graph represents mesh updates (removals and/or insertions of triangles that change the representation of a mesh region). The arcs of the graph represent relations among updates. An arc $a = (n_1, n_2)$ exists if a nonempty subset of the triangles introduced by the update represented by node n_1 is removed by the update represented by n_2. Selectively refined meshes can thus be obtained from cuts of this graph. Sweeping the cut forward or backward through the DAG increase or decreases the resolution. Figure 3.4 illustrates the concept.

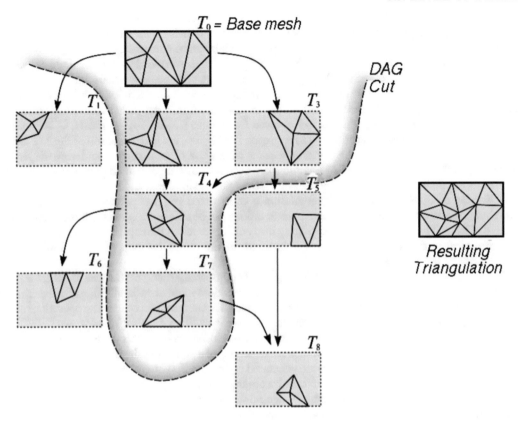

Figure 3.4: **Multi-triangulation.** A sequence of local modification over a mesh is coded as DAG over the mesh fragments T_i; a cut of the DAG defines a conforming triangulation that can obtained by pasting all the mesh fragments above the cut.

Most of the continuous LOD models can be expressed through this framework. Many variations have been proposed. Until recently, the vast majority of view-dependent level-of-detail methods were all based on multi-resolution structures that made decisions at the triangle/vertex primitive level. This kind of approach involves a constant CPU workload for each triangle and makes detail selection the bottleneck in the entire rendering process. This problem is particularly problematic in rasterization because of the increasing CPU/GPU performance gap.

3.3.4   Coarse-Grained Continuous LOD Models

To overcome the detail selection bottleneck and to fully exploit the capabilities of current hardware, it is necessary to select and send batches of geometric primitives to be rendered with just a few CPU instructions. To this end, various GPU oriented multi-resolution structures have been recently proposed. The proposals are based on the idea of moving the granularity of the representation from triangles to triangle patches [CGG$^+$04, YSGM04]. Thus, instead of working directly at the triangle level, the models are first partitioned into blocks that contain many triangles. A multi-resolution structure is then constructed among partitions. By carefully choosing appropriate subdivision structures for the partitioning and managing boundary constraints, hole-free adaptive models can be constructed.

The benefit of these approaches is that the needed per-triangle workload to extract a multi-resolution model reduces by orders of magnitude. The small patches can be preprocessed and optimized off-line for more efficient rendering, and highly efficient retained mode graphics calls can be exploited for caching the current adaptive model in video memory. Recent work has shown that the vast performance increase in CPU/GPU communication results in greatly improved frame rates [CGG+04, YSGM04, CGG+05].

The Batched Multi-Triangulation (MT) approach [CGG+05] is a generalization of the MT framework that works on mesh regions and encompasses a wide class of construction and view-dependent extraction algorithms. In the Batched MT approach, a sequence of coarser and coarser partitions is used to define sets of patches over a massive mesh. The patches can be simplified and merged together to form a well-behaving MT DAG. Once the representation is constructed and dependencies among regions are recorded, view-dependent conforming mesh representations can be efficiently extracted by combining precomputed patches. The coarse grained partitioning into patches provides the capability to not only perform coarse-grained view-dependent (or selective) refinement of the model but also to be used for visibility computations and out-of-core rendering.

The original article [CGG+05] explains a general approach for doing this as well as specializations based on Voronoi partitioning. The basic idea is to use hierarchical volumetric subdivision schemes that are able to create conforming volumetric meshes that support local refinement and coarsening operations and to then exploit them for adaptively refining an embedded surface mesh. The *Adaptive TetraPuzzles* [CGG+04] (ATP) approach partitions the mesh according to a conformal hierarchy of tetrahedra. The partitioning structure is a binary forest of tetrahedra, whose roots correspond to six tetrahedra constructed around a major box diagonal and whose other nodes are generated by tetrahedron bisection. This operation consists of replacing a tetrahedron σ with the two tetrahedra obtained by splitting σ at the midpoint of its longest edge by the plane passing through the longest-edge midpoint and the opposite edge in σ. To guarantee that a conforming tetrahedral mesh is always generated after a bisection, all the tetrahedra sharing their longest edge with σ are split at the same time. Such a cluster of tetrahedra is called *diamond*.

The hierarchy of the tetrahedra structure has the important property that, by selectively refining or coarsening it on a diamond-by-diamond basis, it is possible to extract conforming variable resolution volumetric mesh representations. This property is exploited to construct a level-of-detail structure for the surface of the input model. The basic idea is to use the tetrahedral structure to construct a hierarchy of surface representations. The construction method first partitions the input model triangles among the leaf tetrahedra and then recursively associates a simplification (up to a given triangle count) of the portion of the mesh contained in its two children and all the information required for evaluating view dependent errors to each nonleaf tetrahedron.

To guarantee that each mesh composed of a collection of small patches arranged as a correct hierarchy of tetrahedra generates a globally conforming surface triangulation, simplification is performed on a diamond-by-diamond level. The diamond mesh is simplified as a whole while locking the vertices lying on the diamond's boundary (see Fig. 3.6). Once the diamond is simplified, data is redistributed into the tetrahedra that compose it. By keeping the external boundary fixed, one ensures that the each atomic operation (diamond coarsening and refining) keeps the mesh conformant. It is worth mentioning that, unlike other hierarchical simplification approaches [Hop98, Pri00], these constraints have little effect on overall simplification quality since boundaries are not maintained across hierarchy levels. Similar structures have been presented for 2D domains and have been used for terrain visualization [CGG+03a, CGG+03b], streaming [BGMP07], and compression [GMC+06]. In this case, the partitioning structure is the well-known right triangle hierarchy.

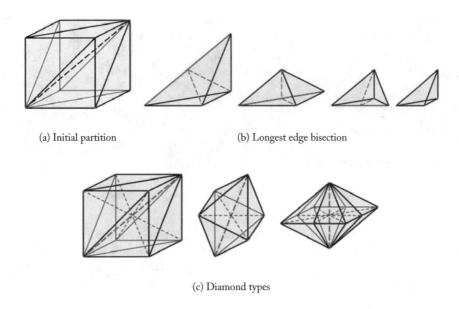

(a) Initial partition   (b) Longest edge bisection

(c) Diamond types

Figure 3.5: **Hierarchy of tetrahedra for space partitioning.** The longest edge is highlighted in red, while next level edges are light-dashed.

At run-time, selective refinement queries based on projected error estimation are performed on the external memory tetrahedron hierarchy. The queries rapidly produce view-dependent continuous mesh representations by combining precomputed patches contained in tetrahedra. An MT DAG with the current representation is maintained and selectively refined or coarsened based on view-dependent error estimates. This operation can be efficiently performed in a time-critical fashion using a dual-queue approach for selectively coarsening and refining [DWS+97].

The *Quick-VDR* [YSGM04] approach also exploits a dynamic LOD representation based on coarse mesh partitioning. In contrast to the above spatial subdivision approaches, the partitioning is constructed by first subdividing the mesh into regions and then applying a multilevel k-way partitioning scheme for irregular graphs [KK98]. They also presented improved algorithms to achieve higher level of simplification along the boundaries of these regions without introducing cracks. Moreover, instead of representing each cluster with static meshes, progressive meshes are used to support fine-grained local refinement and to compute an error-bounded simplification of each cluster at runtime. The major benefit of such a representation is its ability to provide efficient both fine-grained and coarse-grained refinement. The representation adds the possibility of having smooth transitions between different LODs. A major problem with this method is relatively low GPU vertex cache utilization during rendering dynamic LODs compared to rendering static LODs. Later, this approach was combined with cache-oblivious layouts to improve the cache efficiency [YLPM05].

The coarse-grained techniques discussed in this section have proven to be able to maximally exploit the capabilities of current graphics boards. They are therefore able to render very dense meshes with a very high quality at interactive rates. Figure 3.7 shows an example of the extremely detailed representations that can be inspected in real-time using these techniques.

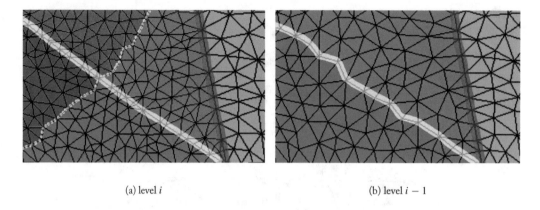

(a) level i (b) level $i - 1$

Figure 3.6:   **Generating conforming triangulations.** The four patches at the left of Fig. 3.6(a) are part of the same diamond, and are simplified into the two patches at the left of Fig. 3.6(b) when coarsening the mesh. The generation of a conforming triangulation is ensured by locking the vertices shared with the neighboring diamond (highlighted in red), and by consistently simplifying the vertices shared by different patches in the diamond (highlighted in yellow).

3.4   DISCUSSION

Level-of-detail and visibility culling techniques are fundamental capabilities for massive model rendering applications. It is important to note that, in general, the lack of one of these techniques limits the theoretical scalability of an application. However, massive models arise from a number of different domains, and the relative importance of LOD management and visibility culling depends on the extent of the geometry variation, the appearance, and the depth complexity characteristics of the models. For instance, typical 3D scanned models and terrain models tend to be extremely dense meshes with very low depth complexity and favors pure LOD techniques. Architectural and engineering CAD models tend to combine complicated geometry and appearance with a large depth complexity and force applications to deal with visibility culling problems.

   The geometric simplification and level of detail techniques discussed in this chapter are based on mesh simplification. Mesh simplification can be considered a mature field for which industrial quality solution exist and are well established. However, these methods, which repeatedly merge nearby surface points or mesh vertices based on error minimization considerations, perform best for highly tessellated surfaces that are otherwise relatively smooth and topologically simple. In other cases, it becomes difficult to derive good "average" merged properties. While it is true that there are quite a few algorithms that perform nice topological simplifications with good error bounds (e.g., [EM99]), pure geometric simplification remains hard to apply when the visual appearance of an object depends on resolving the ordering and mutual occlusion of even very close-by surfaces, potentially with different shading properties. For such complex models, visibility preprocessing and model simplification are strictly coupled.

   Few approaches exist that integrate LODs with occlusion culling in both the construction and rendering phases. Notable exceptions are hardly visible sets [ASNB00] and visibility guided simplifica- tion [ZT02]. Both are nonconservative techniques that favor model simplification in areas that are likely to be occluded. Most importantly, the off-line simplification process that generates the multi-resolution hierar-

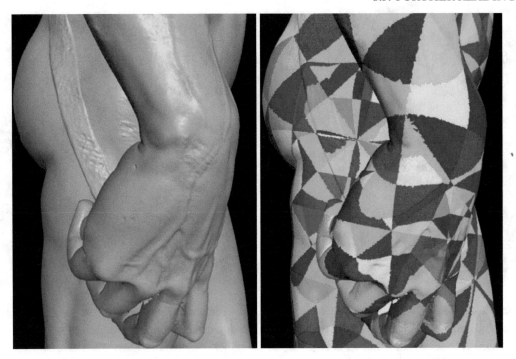

Figure 3.7: David 1mm hand close-up using the ATP [CGG⁺04] technique. Left: model rendered at ±1 pixel screen tolerance with 841 patches and 1172K triangles at 50 fps on a 1280 × 1024 window with 4× Gaussian Multisampling, one positional light and glossy material. Note the very fine geometric and illumination details. Right: mesh partitioning is emphasized using a different per patch color. Images courtesy of CRS4 and ISTI CNR.

chy from which view-dependent levels of detail are dynamically extracted is typically view-independent. The process is either essentially unaware of visibility or just tries to take into account spatial properties at a coarse level to build structures that have good properties for visibility culling [ISGM02, CGG⁺04, YSGM04]. When approximating very complex models, however, resolving the ordering and mutual occlusion of even very small or close-by surfaces that may have shading properties is of primary importance (see Fig. 3.8).

Providing good multi-scale visual approximations of general environments remains an open research problem, and the few solutions proposed so far involve primitives other than triangle meshes for visibility-aware complexity reduction. Chapter 4 focuses on using alternative primitives for real-time rendering.

3.5   FURTHER READING

Complexity reduction has been an integral part of computer graphics for three decades and a number of good surveys exist. Classic overviews of the subject are presented in the 2002 books by Luebke et al. [LRC⁺02], which is devoted to LOD, and of Akenine-Moeller et al. [AMH02], which has sections on visibility, image based rendering, and simplification. A more in-depth treatment of multiresolution methods for geometric objects is provided by the 2005 collection by Dodgson et al. [ND05]. The recent survey by Pajarola and

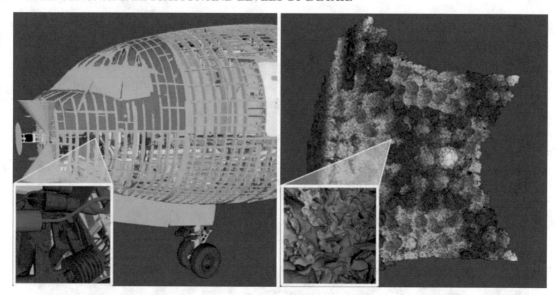

Figure 3.8: **Boeing 777 engine details (left) and isosurface details (right).** These kinds of objects, composed of many loosely connected interweaving detailed parts of complex topological structure, are very hard to simplify effectively using off-line geometric simplification methods that do not take into account visibility. As seen in the insets, a single pixel gets contributions from many mutually occluding colored surfaces. Images courtesy of CRS4.

Gobbetti [PG07] provides a detailed account of recent multiresolution approaches for terrain rendering, while the recent book of Gross et al. [GP07] focuses on point-based graphics.

CHAPTER 4

Alternative Representations

Computing and exploiting filtered representations of details to reduce rendering complexity is the object of simplification and level of details techniques. So far, we have concentrated on methods centered around efficient procedures for simplifying triangle meshes, arranging details in a multiresolution structure, and efficiently extracting them at run-time to realize adaptive rendering. In this chapter, we will discuss how the complexity of the rendering operation can be also reduced by switching to representations other than triangle meshes.

4.1 INTRODUCTION

Geometrically, polygonal meshes are piecewise linear surface approximations, consisting of a collection of polygons pasted along the edges. They are the "common denominator" of most real-time rendering applications because of their efficiency and versatility. Representations other than polygons, however, offer significant potential for massive model visualization.

On one hand, important model classes, such as CAD models, are well described in terms of higher-order geometric curve, surface, and solid primitives. One might thus consider directly rendering them instead of resorting to precomputed tessellations. The potential advantages of such an approach include reducing memory needs and generating smooth surface views at high magnification levels.

On the other hand, in conventional polygon-based computer graphics, models have become so complex that for most views the projection of polygons may be smaller than one pixel in the final image. As a result, many researchers have been investigating alternative, mostly sample-based, scene representations. These representations use sets of points, voxels, or images to accelerate the rendering.

4.2 HIGHER-ORDER REPRESENTATIONS

Mathematical models are required in most 3D mechanical and architectural design environments. The base representations of most CAD systems are based on high-order representations, such as canonical volume and surface definitions or parametric representations of curves, surfaces, and volumes. Such representations allow downstream engineering and manufacturing analysis to be based on the most accurate design representation possible.

In most massive-model rendering systems, the primitive 3D entities used for rendering are statically pre-computed from these higher-order geometric representations. The pre-computation (often called tessellation) results in a fixed number of 3D primitives (generally polygons that are further decomposed into triangle meshes) that are invariant to zoom. The side effect is that edges become visible at high zoom levels if the tessellation is too coarse.

Both raster-based and ray tracing rendering approaches work directly and efficiently with low-order primitives. High-order primitives are generally tessellated into triangles or into intermediate forms in a preprocessing step.

Direct rendering of the high-order primitives is generally too slow to sustain interactive performance even for relatively small dataset sizes. There have been efforts to integrate high-order primitives into the rendering pipeline since the early 1970's [Gol81]. This kind of work has progressed substantially, and recent approaches are using programming techniques on GPU hardware to increase performance. In particular, a number of authors have focused on devising efficient methods for raycasting quadrics, cubics, and quartics on the GPU [dL04, LB06, TCM06, SWBG06, dLP07], and [KKM07] introduced a method for direct evaluation of NURBS surfaces on the GPU. Even when using specialized hardware, however, current systems do not match the performance of rendering from precomputed meshes. Since the performance of multi-core CPUs and programmable graphics systems continues to grow, it is reasonable to expect that in the near future moderately complex models could be rendered in real-time. This is particularly important for interactive modeling applications, where manipulation of the original parametric data is important.

Intermediate forms have been introduced to deal with the performance issues of dynamic tessellation of high-order definitions. Subdivision surfaces exemplify intermediate forms. In general, some sort of preprocessing is needed to fit the subdivision surfaces to high-order primitives. A second set of algorithms must then be developed to re-tessellate the surfaces on a frame-by-frame basis.

Use of intermediate forms relies on processing on a per-frame basis to determine a new set of polygons to send to either raster or ray tracing algorithms. Because the subdivision surface intermediate form has been generated through pre-processing, much of the complexity (e.g., holes, multi-patch surfaces) that slows high-order primitive rendering down has been replaced with simpler surface forms. The key to success is that successive refinement of the surface is guaranteed to stay on the surface.

There have been numerous publications on subdivision surfaces. Catmull-Clark [CC78] and Doo-Sabin [DS78] represent early examples, and subdivision surfaces have appeared in visualization software since the mid-1980's.

Both high- and intermediate-order display forms have two advantages over low-order primitives: smoother appearance at higher magnification levels and generally smaller data set sizes. In terms of raw visualization speed, however, high-order representations cannot currently be processed efficiently enough to be competitive with massive models formed from triangle meshes. It should also be noted that using higher-order primitives alone does not fully solve the scalability problem of massive model renderers, since at low magnification levels complex models still contain a large number primitives. The definition of a multiresolution representation above the primitive level is therefore required to support view-dependent rendering.

4.3   SAMPLE-BASED REPRESENTATIONS

Sample-based representations are on the opposite end of the spectrum with respect to higher-order representations. They exploit discrete methods to represent complex models with sets of samples.

4.3.1   Point-Based Rendering

Multi-resolution hierarchies of point primitives have recently emerged as a viable alternative to the more traditional mesh refinement methods for interactively inspecting very large geometric models.

One of the major benefits of this approach is its simplicity. Basically, there is no need to explicitly manage and maintain mesh connectivity during both preprocessing and rendering.

The possibility of using points as a rendering primitives was first suggested by Levoy and Whitted [LW85]. They noted that point primitives are more appropriate than triangles for complex, organic shapes with high levels of geometric and appearance detail. Since then, a large body of work has been performed in the area of point-based graphics [GP07].

A point-based geometry representation can be considered a discrete sampling of a continuous surface, which results in 3D positions $\mathbf{p}_i$, that are optionally extended with normal vectors $\mathbf{n}_i$ or auxiliary surface properties, e.g., color or other material properties.

The reconstruction of continuous (i.e., hole-free) images from such a discrete set of surface samples is the major problem faced by point rendering approaches. Methods for closing holes and gaps in-between the samples have to be found. It can be done by image-space reconstruction techniques [GD98] or by object-space resampling. The techniques from the latter category dynamically adjust the sampling rate so that the density of projected points meets the pixel resolution. Dynamic sampling can be used both for rasterization and ray tracing. Since sampling depends on the current viewing parameters, re-sampling has to be done dynamically for each frame, and multi-resolution hierarchies or specialized procedural resampling techniques can be exploited. Examples are bounding sphere hierarchies [RL00c], dynamic sampling of procedural geometries [SD01], the randomized Z-buffer [WFP+01], and the rendering of moving least squares (MLS) surfaces [ABCO+01]. As for polygonal multi-resolution rendering, amortizing over a large number of primitives is essential to maximize rendering speed on current architectures. The highest performance levels are currently attained by coarse-grained approaches.

Coarse-grained refinement for point clouds was introduced by the Layered Point Cloud multiresolution approach [GM04]. The method creates a hierarchy over the samples of the datasets simply by reordering and clustering them into point clouds of approximately constant size arranged in a binary tree. In other words, the final multiresolution model has exactly the same N points of the input model. The input points are grouped into clusters and organized in a level of detail representation. The root of the level of detail tree represents the entire model with a single cloud of $M_0 = M < N$ uniformly distributed samples. The remaining points are equally subdivided between the two subtrees by using a spatial partition, with, again, M uniformly distributed points directly associated to the root of each subtree. Any remaining nodes are redistributed in the children. The leaves are terminal clusters, which are further indivisible and whose size is smaller than the specified limit M.

Variable resolution representations of the models are obtained by defining a *cut* of the hierarchy and merging all nodes above the cut. In this way, each node acts as a *refinement* of a small contiguous region of the parent. This is different from most other hierarchical schemes, where only the leaf nodes of the cut hierarchy are used. Figure 4.1 shows an example of images obtained during real-time visualization of a large scanned model.

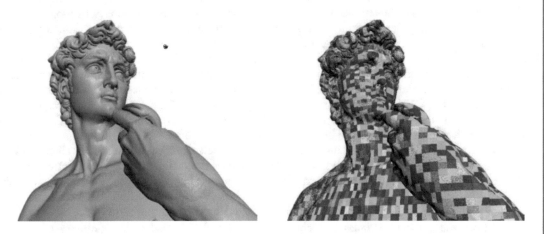

Figure 4.1: David 1 mm visualization using the LPC [GM04] technique. Left: model rendered at ± 1 pixel screen tolerance at 50 fps on a 1280×1024 window on a PC with NVIDIA FX 5800 U. Right: point cloud partitioning is emphasized using a different color per cluster. Images courtesy of CRS4.

Currently, point-based rendering techniques are competitive in terms of rendering performance with mesh-based techniques if one uses simple OpenGL hardware support. Hardware support may be used for point cloud rendering. However, current hardware limits the point-based approach's ability to correctly treat in a single streaming pass texture and transparency and causes aliasing artifacts. Overall, peak performance of high quality techniques based on sophisticated point splatting is currently inferior to the performance of corresponding triangle rasterization approaches because current graphics hardware does not natively support essential point filtering and blending operations. This situation might change in the near future. Novel architectures for hardware-accelerated rendering primitives are currently being introduced [WFH+07].

4.3.2   Volumetric Representations

Sample-based representations are appealing in massive model applications not only for rendering but also to serve as modeling primitives for generating LODs. Classically, they have been used to represent surface elements. More recently, they have been used to model the appearance of small volumetric portions of the environment, which offers advantages in models with very complex geometry.

The *Far Voxels* [GM05] system exploits the programmability and batched rendering performance of current GPUs. Far voxels is based on the idea of moving the grain of the multi-resolution surface model up from points or triangles to small voxel clusters. The voxels represent spatially localized dataset regions with groups of (procedural) graphics primitives. The clusters provide the capability of performing coarse-grained view-dependent refinement of the model and are also used for on-line visibility culling and out-of-core rendering.

Figure 4.2 provides an overview of the approach. To generate the clusters, the model is hierarchically partitioned with a kd-tree. Leaf nodes partition full resolution data into fixed triangle count chunks, while inner nodes are discretized into a fixed number of cubical voxels arranged in a regular grid.

The far voxels method assumes that each inner node is always viewed from the outside and at a distance sufficient to project each voxel to a very small screen area (e.g., one pixel or less). Under this condition, a voxel always subtends a very small viewing angle, and a purely direction dependent representation of shading information is thus sufficient to produce accurate visual approximations of its projection.

The method employs a visibility aware sampling and reconstruction technique to construct a view-dependent voxel representation. The first step acquires a set of shading information samples by ray casting the original model from a large number of appropriately chosen viewing positions. Each sample associates a reflectance and a normal to a particular voxel observation direction. The next step then compresses the samples into an analytical form. The analytical form can be compactly encoded and rapidly evaluated at run-time on the GPU to compute voxel shading given a view direction and light parameters. The analytical form is found by fitting the samples to simple parameterized shader models. The algorithm chooses the shader that provides the best approximation. Each unique shader contains a function that returns a color attenuation given its internal parameters, a view direction $\mathbf{v}$, and a light direction $\mathbf{l}$, i.e., $\text{Shader}_i(\mathbf{v}, \mathbf{l}) = BRDF_i(\mathbf{v}, \mathbf{l}) \max(\mathbf{n}(\mathbf{v}) \cdot \mathbf{l}), 0)$, where $\mathbf{n}(\mathbf{v})$ is the surface normal seen from $\mathbf{v}$. Instead of relying on a single general purpose shader, the Far Voxels approach assumes that a small number of shader classes can be used to model and accelerate common situations.

The shader selected for a particular voxel is found by constructing an instance of each shader class k from the gathered samples and choosing the one that provides the minimum error $\varepsilon^{(k)} = \sum_i \sum_j \left(BRDF_i^{\text{(sampled)}}(\mathbf{v}_i, \mathbf{l}_j) \max(\mathbf{n}_i \cdot \mathbf{l}_j, 0) - \text{Shader}^{(k)}(\mathbf{v}_i, \mathbf{l}_j) \right)^2$. At rendering time, the voxel representation, rendered as point primitives, is refined and rendered in front-to-back order, exploiting vertex shaders accelerated by GPU.

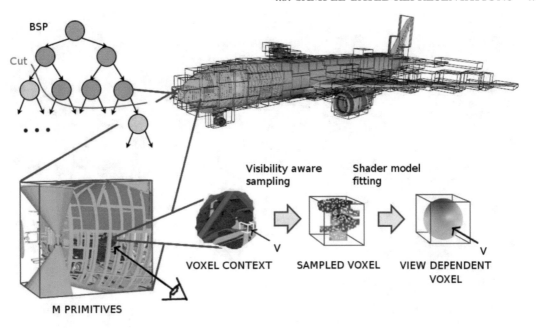

BSP

Cut

Visibility aware
sampling

Shader model
fitting

V

VOXEL CONTEXT   SAMPLED VOXEL   VIEW DEPENDENT
VOXEL

V

M PRIMITIVES

Figure 4.2:   Far voxels overview. The model is hierarchically partitioned with a kd-tree. Leaf nodes are rendered using the original triangles, while inner nodes are approximated using view-dependent voxels.

The resulting technique has proven to be fully adaptive and applicable to a wide range of model classes that contain highly detailed colored objects composed of many loosely connected interweaving detailed parts of complex topological structure (see Fig. 4.3). The major drawbacks of far voxels are the large preprocessing costs and the aliasing and transparency handling problems introduced by the point splatting approach.

4.3.3   Sample-Based LODs for Ray Tracing

Although often neglected, finding good LOD representations is also important for ray tracing systems. In the absence of a suitable LOD representation, the working set size of ray tracing can be very high. When the working set becomes bigger than available memory, paging and disk reads significantly degrade ray tracing performance. Since the leaf nodes of a hierarchy for ray tracing are individual triangles, memory overhead becomes a major issue. To address this issue, the *R-LOD* [YLM06] system has introduced a LOD representation for ray tracing that is tightly integrated with kd-trees. Specifically, an R-LOD consists of a plane with material attributes (e.g., color), which is a drastic simplification of the descendant triangles contained in an inner node of the kd-tree, as shown in Fig. 4.4. In this way, the approach is similar to one of the shaders employed by the Far Voxels system.

Each R-LOD is associated with a surface deviation error, which is used to quantify the projected screen space error at runtime. If an R-LOD representation of a kd-node has enough resolution for a ray according to an LOD metric, further hierarchy traversal for ray-triangle intersection tests stops and performs ray-LOD intersection tests. The method has the disadvantage that it does not provide the complete set of

Figure 4.3: **Far Voxels [GM05] rendering example.** A 1.2 billion triangles scene interactively inspected on a large scale stereoscopic display driven by single PC, which renders two 1024 × 768 images per frame with a 1 pixel tolerance. Image courtesy of CRS4.

LOD solutions for arbitrary rays, especially for nonlinear transformations, such as refractions and reflections off of curved surfaces.

4.4   IMAGE-BASED METHODS

In geometry-based rendering, the visible component of the world is the union of two elements: the geometric description of the objects and the color and lighting conditions. A different approach treats the virtual world as an infinite collection of 2D images taken at different position, orientation, and time. Such a collection of images yields a *plenoptic function*, i.e., a function that returns the color perceived from a specified eye position, view orientation, and time. The goal of *image-based rendering* (IBR) is to generate images by resampling the plenoptic function given the view parameters [MB95]. With this approach, a complex environment can be theoretically represented by a series of images rather than a full three-dimensional model. In this way, rendering complexity is separated from the geometric model complexity of the scene.

Resampling the plenoptic function to generate novel views is difficult, however. The function is five-dimensional and it is necessary to generate enough samples to avoid aliasing. In the general case, a fully

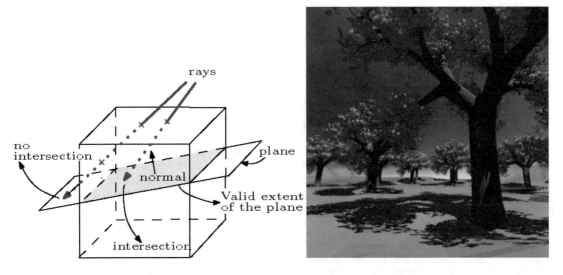

Figure 4.4: R-LOD Representation. A R-LOD consists of a plane with material attributes. It serves as a drastic simplification of triangle primitives contained in the bounding box of the sub-tree of a kd-tree node. Its extent is implicitly given by its containing kd-node. The plane representation makes the intersection between a ray and a R-LOD very efficient and results in a compact representation. On the right, a forest model consisting of 32 M triangles is rendered with a 2×2 supersampling and 4 pixels of error at 1.6fps when using R-LODs, a 5 times improvements with respect to a single resolution approach. Images courtesy of University of North Carolina.

IBR approach is also impractical because of the sheer amount of data required for a full dense light field encoding of a scene. Restricted solutions have been proposed that reduce the complexity of the problem by imposing constraints on viewer motion or by compensating for the aliasing effect by using additional geometric information.

4.4.1   Image-Based Rendering without Geometry

Since compensation for aliasing effects is impossible without additional geometric information, either the sampling must be very dense [GGSC96, LH96], which is impractical for large scenes, or the possible viewer motion must be restricted. The second case has produced some practical approaches for navigating in complex virtual environments.

If the plenoptic function is only constructed for a single point in space, then its dimensionality is reduced from 5 to 2 because there is only angular dependency. This principle is used in environment mapping. In this approach, the view of the environment from a fixed position is represented by a 2D texture map and exploited in spherical or cylindrical panorama systems such as Quicktime-VR [Che95]. Similar constraints can be applied if the user is constrained to move along predefined paths. In that case, the mapping becomes 3D. The user can walk and look around interactively [Lip80]. These approaches are viable only if constrained viewer motion is acceptable.

4.4.2   Image-Based Rendering with Geometry Compensation

In the last decade, a set of successful hybrid techniques have been proposed to accelerate the rendering of portions of a complex scene by replacing well-defined portions of the images with complex, texture-mapped geometry. In most cases, the basic idea is to use a geometry-based approach for near objects and then switch to a radically different image-based representation, called an *impostor*, for distant objects that have a small, slowly changing on-screen projection.

The *billboard*, a textured planar polygon whose orientation changes to always face the viewer, is the most well-known image-based representation. Billboards are used for replacing geometric representations of objects that have a rough cylindric symmetry, like a tree.

Another application of IBR is in environments which are naturally subdivided in cells with reduced mutual visibility. A typical example is the inside of a building, where adjacent rooms can be connected by doors or windows. If the observer is in a room, he or she can see the inside of the adjacent cells only through those openings. This feature can be exploited in visibility culling. All geometry outside the perspective formed by the observer position and the opening can be disregarded. If the observer is not too close to the opening and/or the opening is not too wide, it makes sense to put a texture on the opening instead of displaying the geometry. *portal textures* [AL97] introduced the concept.

These simple approaches are limited. A single texture provides the correct view of the scene only from the point where it has been sampled and nowhere else. This leads to artifacts when the observer moves. For these reasons, a number of authors have proposed more elaborate solutions to incorporate parallax effects, such as *textured depth meshes* [SDB97], in which textures are triangulated and a depth value is associated to each vertex, and *layered depth images* [SGHS98], which store all the intersections of the view ray with the scene for each pixel.

In the *Textured Depth Mesh* approach [SDB97, WM03], textures are triangulated and a depth value is associated to each vertex. The result is an image that is not just a discrete grid of points but a continuous surface, which enables a simple simulation of parallax effects.

In [DSSD99], the disocclusion error generated by associating points that belong to different surfaces to the same impostor is estimated, and objects are grouped in a way that minimizes the error. In [JWS02, JW02], a collection of slices at increasing distance are used as impostors that form a *layered environment-map impostor*. Impostors can be based on *Layered Depth Images*. These impostors store all the intersections of the view ray with the scene for each pixel. The concept was introduced in [SGHS98]. The extension in [WWS01a] replaces the regular sampling of LDIs with a more general adaptive point sampling of the geometry.

These techniques, introduced a decade ago, are enjoying a renewed interest, because of the evolution of graphics hardware. The new hardware is more and more programmable and oriented toward massively parallel rasterization. The hardware evolution also blurs the boundary between geometry-based and image-based representations because more and more geometric information is encoded in the various texture-based representation to increase rendering fidelity.

The techniques used for rendering impostors are strictly related to the issue of height field ray tracing and displacement mapping techniques [Coo84]. A number of specialized hardware accelerated techniques have been recently presented (e.g., *relief mapping* [OBM00] and *view-dependent displacement mapping* [WWT+03]).

The Relief Mapping approach was introduced in [OBM00]. Relief Mapping finds the final position of an orthogonally displaced texel over a given flat texture. An approach to the problem of rendering generalized displacement mapped surfaces by GPU ray casting was proposed in [WWT+03, WTL+04]. In these methods, the results of all possible ray intersection queries within a three-dimensional volume are precomputed and stored in a compressed five-dimensional map. While general, this approach incurs a substantial storage overhead.

Other generalizations include replacing the orthogonal displacement with a more general inverse perspective [BD06] and handling self-shadowing in general meshes [POC05].

A recent evolution of these methods is the BlockMap [CDBG+07]. A Block Map compactly represents a set of textured vertical prisms with a bounded on-screen footprint in a single texture. The texture replaces a set of buildings in city rendering applications (see Fig. 4.5). In many ways, the BlockMap representation is more similar to LOD than to impostor approaches. The BlockMap provides a view-independent, simplified representation of the original textured geometry, provides full support to visibility queries, and, when built into a hierarchy, offers multi-resolution adaptability.

Encoding shape and appearance into a texture is also the goal of geometry images [GGH02], which enable the powerful GPU rasterization architecture to process geometry in addition to images. Finally, there have been a few techniques that these image-based rendering techniques to massive models [CDBG+07, WM03, ACW+99].

Figure 4.5: BlockMaps for large scale urban model rendering. A reconstruction of the city of Paris created from 80,414 building outlines and 19.6G texels of facade textures is inspected in real-time using the BlockMap approach [CDBG+07]. Images courtesy of CRS4 and ISTI CNR.

4.5   DISCUSSION

In this chapter, we have shown how the complexity of the rendering operation can be reduced by employing representations other than triangle meshes. To date, the different solutions developed have advantages and disadvantages and no single best representation exists in terms of storage, computational, and implementation costs. This means that selecting the most appropriate solution is scene and application dependent.

Mesh-based representation are an all-round kind of model that typically provides "reasonable" solutions for a wide variety of situations. A number of systems use them as the sole representation. Voxel- and point-based techniques can also be considered general enough to be used as the sole primitive in a general purpose rendering system. They are, however, more performance for densely sampled models, and current general purpose solutions tend to combine them with polygons for representing the finer detail levels. The other sample-based techniques reviewed in this chapter are typically restricted to particular types of applications (e.g., constrained panoramic viewing), or used to represent well-defined portion of a scene otherwise modeled with more general techniques. Higher-order primitives are currently very rarely used in real-time massive model renderers, but, as hardware progresses, we can expect to see them appear in the near future at least to offer a better support to CAD models visualization.

Today, no universal system exists that can handle all massive models application scenarios. In most cases, the different solutions illustrated in the previous sections must be carefully mixed and matched in a single hybrid but coherent system able to balance the competing requirements of realism and frame rates.

4.6   FURTHER READING

A classic overview of the subject of using alternative representations to speed-up real-time graphics systems is the book by Akenine-Moeller et al. [AMH02]. There are sections on visibility, image-based rendering, and simplification. The recent book of Gross et al. [GP07] provides an in-depth treatment of point-based graphics.

CHAPTER 5

Cache-Coherent Data Management

A major trend on current commodity hardware is the widening gap between data access speed and data computation speed. Data access takes ever-increasing amounts of time in many applications and, therefore, becomes the major bottleneck. Given this trend, it is critical to design efficient data management algorithms that can reduce data access time. In this section, cache-coherent layout techniques that can reduce the number of misses that occur while accessing data are discussed.

5.1 INTRODUCTION

Over the last decade, advances in model acquisition, computer-aided design (CAD), and simulation technologies have resulted in massive datasets containing complex geometric models. Examples include complex CAD models, scanned urban data, and various scientific simulation data. The massive datasets consume gigabytes and even terabytes of storage.

This becomes an issue as the gap among processor speed, main memory, and secondary speed has widened. For example, CPU performance has increased 60% per year for nearly two decades. On the other hand, main memory and disk access time only decreased by 7–10% per year during the same period [RW94, PAC+97]. A relative performance gap between CPU performance and access time to DRAM is shown in Fig. 5.1. As a result, system architectures increasingly use caches and memory hierarchies to avoid memory latency. The access times of different levels in a memory hierarchy typically vary by orders of magnitude. In some cases, the running time of a program is as much a function of its cache access pattern and efficiency as it is of operation count [FLPR99, SCD02].

There are increasing demands to interactively visualize and analyze these complex and massive geometric data sets to extract meaningful information, and scientific discoveries. Given the high model complexity, many traditional algorithms are unable to provide the real-time performance needed for effective interaction. Significant research is needed to design algorithms that can efficiently handle these massive geometric data sets and overcome the increasing gap between data access and computation.

There are increasing demands to process geometric models on different types of computation devices, including PDAs and cell phones. Since those devices have relatively small main memory, there are several performance issues in processing even relatively small models. In addition, these devices typically rely on network data transmission to retrieve the geometric data. Therefore, it is highly desirable to reduce bandwidth requirements as well.

The goal is to design scalable graphics and geometric algorithms that can process large meshes for a wide variety of applications. Examples include rendering and collision detection. Two data reduction techniques for scalable rendering algorithms have been covered in previous chapters: visibility culling techniques in Ch. 2 and simplification methods in Chs. 3 and 4. However, even after performing visibility culling and simplification techniques, it is likely that the number of primitives (e.g., triangles) in a model will exceed CPU memory. Moreover, visibility culling and simplification methods typically require hierarchical culling data structures and multi-resolution representations, which store more data to complement the original mesh. Therefore, it is likely to increase data access times even more.

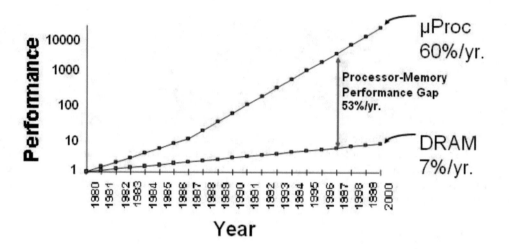

Figure 5.1: **Relative Performance Gap between CPU Processing Power and Access Time to Main Memory:** The CPU performance has increased 60% per year for almost two decades. During the same period, the access time for main memory consisting of DRAM only decreased by 7-10% per year. The graph shown is excerpted from a talk by Trishul Chilimbi.

At a high level, there are two standard techniques used to reduce the data access times.

1. **Computation Reordering.** This method reorders the computation in order to improve cache utilization during execution. This is performed using compiler optimizations, algorithm redesign, or application specific hand-tuning.

2. **Data Layout Optimization.** This technique reconfigures the underlying data layout in order to reduce the number of cache misses. Typically, this is achieved by matching the layout to the data access pattern of an algorithm.

The target is the same: reduce the number of cache misses. The difference is that computation reordering re-orders the access pattern of the runtime algorithm and data layout optimization re-orders the underlying data.

Computational re-ordering techniques in computer graphics and visualization have been active researched early in 2000's. Most techniques can be classified as out-of-core techniques because they require a disk block size that maps portions of the disk into main memory. Out-of-core techniques attempt to minimize the number of disk I/O operations by redesigning algorithms or underlying data structures and keeping the working set size less then the available main memory size. This approach is necessary because disk I/O operations are very expensive compared to other operations.

In this chapter we focus on data layout optimization of large meshes to improve cache coherence of a mesh. A triangle mesh is represented by a set of vertices and triangles. Therefore, the problem is simply to compute separate layouts of vertices and triangles that have the lowest random access data time.

Organization. The rest of the chapter is organized as follows. A brief survey of various cache-coherent layouts is given in Sec. 5.2. Section 5.3 gives an overview of data layout optimization methods. An efficient layout construction method is explained in Sec. 5.4 and is followed by a specialized layout technique for hierarchies (e.g., bounding volume hierarchies) in Sec. 5.5. Different applications that can benefit from the layout techniques are then discussed. Section 5.7 concludes the chapter.

5.2   SURVEY OF CACHE-COHERENT ALGORITHMS

Cache-coherent algorithms have received considerable attention over the last two decades in various fields of computer science, including theoretical computer science, architecture, and compiler literature. These algorithms include models of cache behavior [Vit01] and compiler optimizations based on tiling, strip-mining, and loop interchanging; all of these algorithms have been shown to reduce cache misses [CM95]. Two major reduction techniques that reduce data access time are covered: computational reordering and data layout optimization. The focus is on data layout optimization methods for graphs, meshes, and hierarchies.

5.2.1   Computational Reordering

Computational re-ordering methods attempt to reduce the number of cache misses of an algorithm by re-ordering computational operations of the algorithm. This is typically performed using automatic compiler optimizations or manually re-designing the algorithm. Computation re-ordering techniques can be classified as *cache-aware* or *cache-oblivious*. There has been a considerable amount of literature on designing cache-aware and cache-oblivious computational re-ordering algorithms for a wide variety of algorithms, including numerical programs, sorting, geometric computations, matrix multiplication, and graph algorithms. Good surveys are available in [ABF04, Vit01].

Figure 5.2: Coal Fire Power Plant: This environment consists of over 12 million triangles and 1200 objects. This model has very irregular distribution of geometry and irregular triangular shapes.

Cache-aware techniques.   There is a significant performance difference when data is accessed from register, L1 cache, main memory, or disk. Often, data access time from disk to main memory or main memory to cache is the major performance bottleneck of various applications. Therefore, it is natural to consider cache information when designing specific algorithms. Cache-aware algorithms directly use knowledge of cache parameters, such as cache block size. Cache-aware methods have been designed for external sorting, searching, linear algebra, computational geometry, and combinatorial problems of graphs [Vit01]. Cache-aware algorithms are also known as external or out-of-core algorithms.

Cache-oblivious techniques.   Unlike cache-aware methods, cache-oblivious algorithms attempt to reduce the data access time without prior knowledge of specific cache parameters [FLPR99]. Instead, cache-oblivious methods are designed to reduce the number of cache misses for all possible cache parameters. Therefore, cache-oblivious methods have the potential to improve performance across all levels of memory hierarchies. However, cache-oblivious methods are likely to perform worse than cache-aware methods that are optimized to specific cache parameters.

Out-of-core mesh processing.   Out-of-core algorithms are designed to handle massive datasets on computers with finite memory. A recent survey of these algorithms and their applications is given in [SCC^{+}02]. Most out-of-core techniques depend on cache block size. Some techniques propose efficient disk layouts that reduce the number of disk accesses and the transfer time needed to load the data. Other algorithms use pre-fetching techniques based on spatial and temporal coherence. These algorithms have been used for model simplification [CMRS03], interactive display of large datasets composed of point primitives [RL00a], polygons [CKS03, YSGM04], mesh compression [IG03], and collision detection [FNB03, WLML99].

5.2.2   Data Layout Optimization of Meshes and Graphs
Data layout optimization methods re-order the underlying data to reduce the number of cache misses. This is mainly because that the order of data elements can have a major impact on runtime performance of the algorithm. Therefore, there has been considerable effort to compute cache-coherent layouts by matching the anticipated access pattern of the algorithm to disk storage layout. This section covers graph and matrix layouts, rendering sequences for computer graphics, processing sequences, and space-filling curves.

Graph and matrix layouts.   Graph and matrix layout problems are combinatorial optimization problems. The main goal is to find a linear layout of data elements of a graph or matrix that minimizes a specific objective function. A well-known minimization problem is minimum linear arrangement (MLA). The problem is addressed by computing a layout that minimizes the sum of edge lengths, i.e., index differences of adjacent vertices. The MLA problem is known to be NP-hard, and the decisive version is NP-complete [GJS76]. However, its importance in many applications has inspired a wide variety of approximations based on heuristics. A good example is spectral sequencing [JM92], which minimizes the sum of squared index differences of edges. However, there has been no evidence that MLA can reduce the number of cache misses of runtime applications operating on the graphs. There are other objective functions, such as bandwidth (maximum edge length) and profile (sum of maximum per-vertex edge length). This work has been widely studied and an extensive survey is available [DPS02]. However, these layouts do not necessarily reduce the number of cache misses.

Rendering sequences. Modern GPUs maintain a small vertex buffer to reuse recently accessed vertices. In order to maximize the benefits of vertex buffers for fast rendering, triangle reordering has been employed. This approach was pioneered by Deering [Dee95]. The result is called a triangle strip or a rendering sequence. Hoppe [Hop99a] casts triangle reordering as a discrete optimization problem with a cost function that depends on a specific vertex buffer size that can be computed as a preprocess. If a triangle strip is computed on the fly using view-dependent simplification or other geometric operations, the rendering sequences need to be efficiently recomputed to maintain high rendering throughput and fast updates. Several techniques improve the rendering performance of view-dependent algorithms by computing rendering sequences not tailored to a particular cache size [BG02, KBG02, DGBGP05, YLPM05].

Processing sequences. Isenburg et al. [ILGS03a] proposed processing sequences as an extension of rendering sequences to large-data processing. A processing sequence represents an indexed mesh as interleaved triangles and vertices that can be streamed through main memory [IL05]. Global mesh access is restricted to a fixed traversal order; only localized random access to the buffered part of the mesh is supported as it streams through memory. This representation is mostly useful for offline applications (e.g., simplification and compression) that can adapt their computations to the fixed ordering.

Space-filling curves. Many algorithms use space-filling curves [Sag94] to compute cache-friendly layouts of volumetric grids or height fields. These layouts are widely used to improve the performance of image processing [VdMG91] and terrain or volume visualization [LP01, PF01]. A standard method of constructing a mesh layout based on space-filling curves is to embed the mesh or geometric object in a uniform structure that contains the space-filling curve. An example of Z-curve on a 4×4 uniform grid is shown in Fig. 5.3.

Gotsman and Lindenbaum investigated the spatial locality of space-filling curves [GL96]. Motivated by searching and sorting applications, Wierum [Wie02] proposed using a logarithmic measure of edge length for analyzing space-filling curve layouts of regular grids. Space-filling curve constructed in a uniform grid does not work well for meshes that have an irregular distribution of geometric primitives. Therefore, these algorithms have been used for objects or meshes with a regular structure (e.g., images and height fields).

5.2.3 Cache-Coherent Layouts of Hierarchies

The impact of different layouts of tree structures has been widely studied. There is a considerable amount of work on cache-coherent layouts of tree-based representations. This includes work on accelerating search queries, which traverse the tree from the root node to descendant nodes. Given the cache parameters, Gil and Itai [GI99] considered cache-coherent layouts given cache parameters as an optimization problem. They proposed a dynamic programming algorithm to minimize the number of cache misses during traversals of search queries. However, the computed layout may not be storage efficient and the size of a layout of a tree can be two times bigger than its original tree size. Alstrup [ABD+03] proposed a method to compute cache-oblivious layouts of search trees by recursively partitioning the trees.

Bounding volume hierarchies. Bounding volume hierarchies (BVHs) have been widely used in many different geometric applications including visibility culling, collision detection, and ray tracing. However, there has been relatively less work on cache-coherent layouts of BVHs. Opcode[1] used a blocking method that

---

[1] http://www.codercorner.com/Opcode.htm

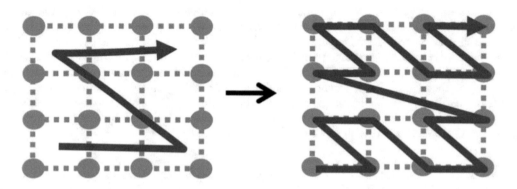

Figure 5.3: Z-curve on a uniform grid: These two Z-curves shows the recursive construction of Z-curves on a uniform grid. The left figure shows a Z-curve on 2×2 virtual uniform grid where each node in the z-curve corresponds 2×2 sub-grid. Then, each node is refined into 2×2 grid and compute a full Z-curve on the 4×4 uniform grid.

merges several bounding volumes nodes together to reduce the number of cache misses. The blocking is a specialized technique based on *van Emde Boas* layout of complete trees [vEB77]. The van Emde Boas layout is computed recursively. Given a complete tree, the tree is partitioned with a horizontal line so that the height of the tree is divided into half. The resulting sub-trees are linearly stored by first placing the root sub-tree followed by other sub-trees from leftmost to rightmost. This process is applied recursively until it reaches a single node of the tree. However, it is not clear whether the van Emde Boas layout minimizes the number of cache misses during traversal of BVHs that may not be balanced or complete trees.

5.3   OVERVIEW OF DATA LAYOUT OPTIMIZATION

This section covers the problem of computing cache-coherent layouts of meshes and hierarchies into data layout optimization. The major goal of optimization is to construct cache-coherent layouts of meshes and hierarchies that have low data access time while reading the layout in a random order at runtime. Performing effective optimization requires a metric that measures the data access time while using the computed layout. The following subjects are covered: an overview of the data layout optimization method, an I/O model, and derivations of optimization metrics based on the I/O model.

Data access graph.   The layout algorithm discussed requires an input data access directed graph, $G = (N, A)$, where N is the set of data and A is a set of directed arcs representing data access pattern between two data nodes. For example, if a runtime application is likely to access a node n_2 right after another node n_1, a directed arc (n_1, n_2) is created in the graph. The data access directed graph represents the data access patterns for data that an application accesses at runtime. For the rest of the discussion, we call the graph a data access graph for the sake of clarify. An example of a data access graph that consists of four nodes is shown in Fig. 5.4. Each arc in the data access graph is assumed to be accessed with equal probability by a runtime application.

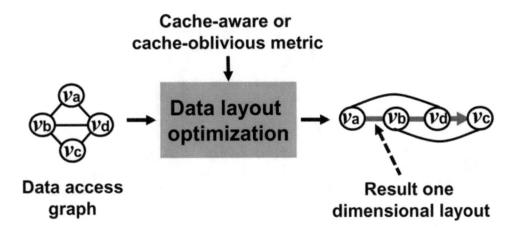

Figure 5.4: **Overview of data layout optimization:** This figure illustrates the overall data layout optimization method. The initial data access graph represents the data access pattern of an application. This is followed by construction of a one-dimensional layout of the graph by solving an optimization problem with a metric that measures the expected number of cache misses.

I/O model. Most modern computers use hierarchies of memory levels. Each level of memory serves as a *cache* for the next level. An example of a memory hierarchy is shown at Fig. 5.5. Memory hierarchies have two main characteristics. First, lower levels are larger in size and farther from the processor and have slower data access times. With typical commodity hardware, L1/L2 caches contain multiple megabytes and are the closest to CPUs. Since these caches are very close to CPU, their cache miss penalty is a few microseconds. Main memory and disk have multiple gigabytes and hundreds of gigabytes, respectively. Their data access time are in the order of tens of microseconds and a few milliseconds, respectively.

Second, data is moved in blocks between different memory levels that contain multiple elements. Typical block sizes of L1/L2 caches are 32 and 64 bytes. Typical page size in main memory is 4KB. Data is initially stored in the slowest memory level, typically disk. A transfer is performed whenever there is a cache miss between two adjacent levels of the memory hierarchy. Due to this *block fetching* mechanism, cache misses can be reduced by storing data elements that are accessed together in the same block. Therefore, the number of cache misses is dependent on the layout of a mesh and the access pattern of the application.

A simple *two-level I/O-model* was defined by Aggarwal and Vitter [AV88]. The model captures the two main characteristics of a memory hierarchy. This model assumes a fast memory called a "cache" that consist of M blocks and a slower infinite memory. The size of each cache block is B; therefore, the total cache size is $M \times B$. Data is transferred consecutively between the levels.

Layout and mesh layout. A layout of a data access graph $G = (N, A)$ is a linear sequence of nodes of the graph. More specifically, a layout of the graph is an one-to-one mapping of nodes to positions, $\varphi : N \to \{1, \ldots, n\}$, where $|N| = n$.

A mesh layout requires computing of two separate layouts: a vertex layout and a triangle layout. The algorithm for computing a mesh vertex layout requires a data access graph where vertices of the mesh correspond to nodes of the graph. Since many geometric applications access the vertices of a mesh

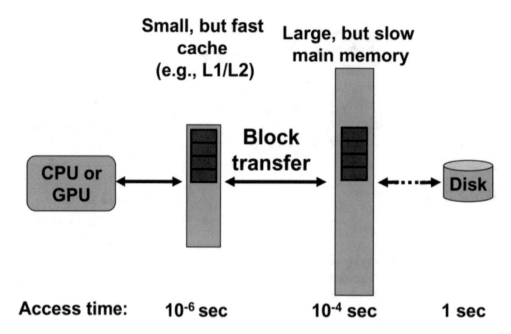

Figure 5.5: **Memory Hierarchy:** This figure shows a memory hierarchy consisting of fast cache, slow main memory, and disk. Lower level has larger space, but has slower data access speed.

by traversing edges of the mesh, edges between two vertices can be used as arcs between two nodes corresponding two vertices in the data access graph. Similarly, a data access graph for a mesh triangle layout can be computed. In this case, each node of the graph corresponds to a triangle of the mesh, and an arc is constructed between two adjacent triangles.

Layout optimization.   The goal is to find a mapping, φ, of a data access graph that minimizes the number of cache misses during access to the graph. This is a layout optimization problem with a metric that measures the number of cache misses during access of the graph. However, it is impossible to precisely compute the number of cache misses at run time before program execution. Therefore, the layout optimization process requires a metric that strongly correlates with the number of cache misses during the runtime application. The *expected* number of cache misses of a graph is approximated by the probability of the number of cache misses during node access. Other properties that the metric should have are simplicity and easy, fast computation. The metric must be very fast since the metric will be used frequently during layout optimization process.

Cache-coherent metrics.   Cache-coherent metrics can be classified into two types: *cache-aware* and *cache-oblivious* metric. A cache-aware metric measures the expected number of cache misses when the cache block size is available. On the other hand, a cache-oblivious metric does not assume a particular block size and accepts variable block sizes. We call these layouts cache-aware and cache-oblivious, respectively. Both cache-aware and cache-coherent metrics measure the expected number of cache misses during random access to data as defined in the data access graph.

Comparison between cache-aware and cache-oblivious metrics.   It is possible that a cache-aware layout optimized with a particular block size is likely to produce better performance than a cache-oblivious layout. However, in the event that a different block size is encountered when using a cache-aware layout construction method, the cache-aware layout is likely to produce sub-optimal performance. If a cache-aware layout can be computed efficiently, it is recommended that a new layout is dynamically re-computed with the new block size. However, the layout construction method discussed in this chapter is not suitable for runtime re-computation of layouts. Moreover, current processing architectures use various levels of memory hierarchies. It is virtually impossible to know all the memory levels that will be involved during data access and consider them during a cache-aware layout construction. On the other hand, cache-oblivious layouts are optimized to handle multiple block sizes. Cache-oblivious layouts provide high performance with various block sizes and can be adapted to different machines with different cache parameters.

5.3.1   Cache-Aware Metric

In this section, cache-aware metrics are derived based on specific computation models. The metrics are used as input to an efficient construction algorithm for cache-aware mesh layouts. The goal is to measure the expected number of cache misses of a layout when accessing a single arc. Since the basic assumption is that arcs are equally likely to be accessed, this measure generalizes to any number of accesses. There are two separate cases of this problem: the cache consists of exactly one block ($M = 1$), and the cache holds multiple blocks ($M > 1$).

5.3.1.1   Single cache block, $M = 1$

Since the cache can only hold one block, a cache miss occurs whenever a node is accessed that is stored in a block different from the cached block. In other words, a cache miss occurs when an arc (n_i, n_j) is traversed and the block containing node n_j is different from the block that holds n_i. Therefore, the expected number of cache misses, $ECM_1^B(\varphi)$, of a layout, φ, for a single-block cache with block size B can be computed as:

$$ECM_1^B(\varphi) = \sum_{\substack{(n_i, n_j) \in A \\ \varphi^B(n_i) \neq \varphi^B(n_j)}} Pr(n_i, n_j)$$

$$= \frac{1}{|A|} \sum_{(n_i, n_j) \in A} S\left(\left|\varphi^B(n_i) - \varphi^B(n_j)\right|\right) , \tag{5.1}$$

where $Pr(n_i, n_j)$ is a probability that a cache miss will occur when block n_j is accessed by traversing the arc (n_i, n_j) and $\varphi^B(n_i) = \lfloor \frac{\varphi(n_i)}{B} \rfloor$ denotes the index of the block in which n_i resides. Otherwise, $S(x)$ is a unit step function such that $S(x) = 1$ if $x > 0$ and $S(x) = 0$.

Simply speaking, $ECM_1^B(\varphi)$ is the number of arcs whose two nodes are stored in different blocks divided by the total number of arcs in the graph.

Layout algorithm.   Constructing a layout optimized for $ECM_1^B(\varphi)$ reduces to a k-way graph partitioning problem. Each directed arc has a constant weight, $\frac{1}{|A|}$. The input graph is partitioned into k different sets of vertices, where $k = \lceil \frac{n}{B} \rceil$. The size of each set should be same as the block size, B, if the number of nodes, n, is a multiple of the block size, B. Otherwise, there will be only one set that does not have enough vertices to fill a block; this set should be stored at the end of the layout.

Since graph partitioning is an NP-hard problem, heuristics are needed for efficiency. One good heuristic is the multi-level graph partitioning algorithm implemented in the METIS library [KK98][2]. Once the directed graph is partitioned, the ordering among blocks and the order of nodes within each block do not matter. Therefore, no specific order within nodes of each block is required.

5.3.1.2   Multiple cache blocks, $M > 1$

The second case is that the cache can hold multiple blocks. In this case, a cache miss can happen only when an arc, (n_i, n_j), that crosses a block boundary is traversed, i.e., $\varphi^B(n_i) = B_i \neq B_j = \varphi^B(n_j)$. However, unlike the case of a single-block cache, block B_j may already be stored in the cache. Therefore, the expected probability, $Pr_{\text{cached}}(B_j)$, that block B_j is cached among the M cache blocks must be computed to predict the expected number of cache misses for a multi-block cache.

In theory, $Pr_{\text{cached}}(B_j)$ can be computed by exhaustively generating all possible access patterns for which block $B(n_j)$ is already cached when n_j is accessed from n_i. Such block access patterns take on the form $(B_j, \ldots, B_i, B_j)$, where the pattern consists of at most M different blocks before B_j is accessed the second time.

Unfortunately, generating all possible block access patterns is prohibitively expensive because of its exponential combinatorial nature. Furthermore, it is not feasible to approximate $Pr_{\text{cached}}(B_j)$ within an error bound without considering a very large number of access patterns. However, it is probable that there is a strong correlation between the numbers of cache misses for a single cache block and multiple cache blocks. To support this conjecture, ten different layouts were computed on a 256-by-256 uniform grid. The number of cache misses incurred in a LRU-based cache was computed during a series of random walk by traversing one of any neighboring nodes. The result showed that the observed number of cache misses with a single block cache has a high linear correlation, 0.9, with the observed number of cache misses when there are a multiple cache blocks ($M = 40$) (see Fig. 5.6).

5.3.2   Cache-Oblivious Metric

This section covers a cache-oblivious metric that measures the expected number of cache misses, $ECM_1(\varphi)$, for varying block sizes.

Block sizes.   In order to derive a cache-oblivious metric that will work well with various levels of memory hierarchies, the block sizes typically employed in memory hierarchies must be analyzed. The first observation is that most block sizes employed in practice have power-of-two bytes (e.g., 32B for L1, 64B for L2, 4KB for disk blocks). Second, the hierarchical relationship between cache levels is often geometric, which suggests that cache block sizes must be optimized at different scales. Therefore, block sizes are used that are geometrically increasing in size (e.g., $1, 2, 4, 8, \ldots$). In fact, block sizes that are linearly increasing to derive a cache-oblivious metric [YL06] are also used. However, the metric with linearly increasing block sizes does not correlate well with the observed number of cache misses, while a metric derived with geometrically increasing block size does. Therefore, this chapter derives a cache-oblivious metric with geometrically increasing block sizes.

The cache-oblivious metric is based on the cache-aware metric with geometrically increasing block sizes. Previously, the cache aware metric was based on a block size B and assumed that only one block can be cached. To derive the cache-oblivious metric, the assumption is made that each block size is equally

---

[2]METIS works only for undirected graph. Directed arcs do not play a role in partitioning since (n_i, n_j) is cut between two different sets and (n_j, n_i) is also cut.

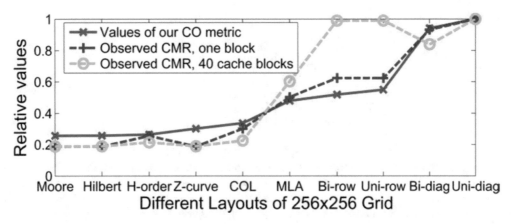

Figure 5.6: **Correlation between Cache Misses and the Computed Metric:** We computed different layouts of a 256 by 256 uniform grid and measured the number of cache misses during random walks on the grid. We found that the cache-oblivious (CO) metric and the observed number of cache misses for a single-block and multi-block ($M = 40$) cache correlated well, with correlation coefficients $R^2 = 0.98$ and $R^2 = 0.79$, respectively. Layouts included are Moore, Hilbert, Z-curve [Sag94], H-order [NRS97], MLA [MP80], the cache-oblivious layout (COL), row-by-row (**row**), and diagonal-by-diagonal (**diag**) layouts. **Uni-** indicates we traverse each row/diagonal from left to right; **Bi-** indicates that we alternate direction. The uni-diagonal layout is the optimal spectral layout. **CMR** indicates cache miss ratio.

likely to happen at runtime. For simplicity, it is further assumes that a layout may start anywhere in the middle of a block with a uniform distribution. This is not unrealistic since a call to obtain memory may return an address anywhere within a memory page or lower level cache block.

Correlated metrics.   Since a particular cache block is employed during runtime, it is practically impossible to correctly measure the exact number of cache misses from the expected number of cache misses reported by the metric. This is mainly caused by the fact that $ECM_1(\varphi)$ is computed by considering all possible cache blocks. Instead, a metric is linearly correlated with the expected number of cache misses. That is, scaling or addition of constants can be factored out of a metric without affecting the correlations. This property is used to derive cache-oblivious metrics that do not measure exactly but correlate well with the expected number of cache misses.

The cache-oblivious metric is expressed in terms of the cache-aware metric, $ECM_1^{2^B}(\varphi)$ where 2^B is a block size and B can be linearly increasing starting from zero. To denote that a layout starts at the kth position in a block, $ECM_1^{2^B}(\varphi)$ is extended to $ECM_1^{2^B}(\varphi, k)$. Then, the cache-oblivious metric ECM_1

can be written as:

$$ECM_1(\varphi) = \sum_{B=1}^{t} \frac{1}{t} \sum_{k=0}^{2^B-1} \frac{1}{2^B} ECM_1^{2^B}(\varphi, k)$$

$$= \frac{1}{t|A|} \sum_{(n_i,n_j)\in A} \sum_{B=1}^{t} \left\{ \sum_{k=0}^{2^B-1} \frac{1}{2^B} S\left(|\varphi^{2^B}(n_i, k) - \varphi^{2^B}(n_j, k)|\right) \right\} \qquad (5.2)$$

where $\varphi^{2^B}(n_i, k) = \lfloor \frac{\varphi(n_i)+k}{2^B} \rfloor$ and t is the maximum number of blocks.

The above equation is very expensive to compute. Therefore, it is not suitable to be used in an optimization method as a metric. Therefore, a few approximations can be used to simplify the equation. Let l be the arc length between two nodes, n_i and n_j, that is, $l = |\varphi(n_i) - \varphi(n_j)|$. The term within brackets in Eq. (5.2) is very expensive to compute. Instead, it is expressed as the probability, $Pr_{cross}(l, 2^B)$, that an arc of length l crosses a block boundary.

The probability is computed in the following manner. Suppose that an arc has an arc length l. The two nodes of the arc are always stored in different blocks when $2^B \le l$, irrespective of where within a block the layout starts. Now consider the case $2^B > l$. There are l different positions for which the arc crosses a block boundary among 2^B different positions at which the layout may start. Therefore, the probability, $Pr_{cross}(l, 2^B)$ that the arc crosses a block boundary is:

$$Pr_{cross}(l, 2^B) = \begin{cases} 1 & (2^B \le l) \\ \frac{l}{2^B} & (2^B > l) \end{cases}. \qquad (5.3)$$

Suppose that c is the number of graph nodes that fit into the smallest block size, $2^0 = 1$ bytes. Then, for a block with a block size 2^B bytes, $c2^B$ nodes fit in a block. Let k be the number of power-of-two byte blocks smaller than the arc length l, that is, k satisfies $c2^k = l$; therefore, $k = \log_2\left(\frac{l}{c}\right)$.

Then, the expected number of cache misses becomes:

$$ECM_1(\varphi) = \frac{1}{t|A|} \sum_{(n_i,n_j)\in A} \int_0^k 1 + \int_k^t \frac{l}{c2^B} dB$$

$$= \frac{1}{t|A|} \sum_{(n_i,n_j)\in A} k + \frac{l}{c}(2^{-t} - 2^{-k}) \qquad (5.4)$$

$$= \frac{1}{t|A|} \sum_{(n_i,n_j)\in A} \log_2\left(\frac{l}{c}\right) + \frac{l}{c2^t} - 1.$$

One can show that the use of integrals instead of sums above introduces negligible error and simplifies the math.

After proper scaling and removal of constants in order to simplify the equation while maintaining the linear correlation, we reach a final cache-oblivious metric.

$$ECM_1(\varphi) \propto \sum_{(n_i,n_j)\in A} \ln\left(|\varphi(n_i) - \varphi(n_j)|\right). \qquad (5.5)$$

For block sizes that follow any power series (i.e., other than powers of two), one can reach the same equation of Eq. (5.5).

The cache-oblivious metric indicates that the probability that a cache miss an arc is increasing sublinearly. This is justified by the fact that once an arc in a layout is long enough to cause a cache miss, lengthening it will not drastically increase the probability of additional cache misses.

Validation.   The correlation between the cache-oblivious metric and the observed number of cache misses have been measured with different layouts of a uniform grid. The number of cache misses are measured when there is only one cache block in the cache during random walks on the layout. The walks include Moore, Hilbert, Z-curve [Sag94], H-order [NRS97], the optimal MLA [MP80], row-by-row, diagonal-by-diagonal layouts of a 256-by-256 uniform grid. There is a strong correlation, 0.98, between the cache-oblivious metric and the observed numbers of cache misses that occurred during random walks on different layouts of the grid (see Fig. 5.6). Although, the cache-oblivious metric assumes that a cache can hold a single block, it is probable that the cache-oblivious metric is correlated with the observed number of cache misses with multiple cache blocks based on the strong correlation observed in the cache-aware case. In order to verify this assumption, the correlation between them was measured by evaluating the correlation between the observed number of cache misses when there are single and multiple blocks during random walks on a uniform grid with different layouts (see Fig. 5.6). There is high correlation, 0.79, between them.

5.4   CACHE-COHERENT LAYOUTS OF MESHES

In this section a layout construction method of triangle meshes will be discussed. The main goal is to find the layout, φ, of the graph $G(N, A)$, such that φ has the minimum value of the cache-oblivious metric, which measures the expected number of cache misses. This is a combinatorial optimization problem that is frequently found in many other graph layout problems [DPS02]. A naive method would be to check all the possible layouts by permuting the nodes of the graph with the cache-oblivious metric. Since there are exponential number of the possible layouts given a graph, it is infeasible to generate all the possible layouts and evaluate each layout with the metric. Simply speaking, finding a globally optimal layout is a NP problem [GJS76] due to the large number of permutations of the set of nodes.

In order to construct a cache-coherent layout of a massive model that minimizes the metric, a heuristic based on *multi-level layout optimization method* is employed. Although, there is no guarantee that the multi-level layout optimization method computes an optimal layout, it has been found that the layout method produce high quality cache-coherent layouts according to various tests. This section explains a multi-level layout method and why the multi-level approach is chosen to construct cache-coherent layouts.

5.4.1   Multi-Level Optimization

The multi-level layout optimization method consists of three main steps: a coarsening step, an ordering step of the coarsest graph, and a refining and local optimization step. First, a series of coarsening operations is computed on the graph. Then, an optimal ordering of nodes of the coarsest graph is constructed by exhaustively permuting all the nodes and choosing the ordering with the minimum value of the cache-oblivious metric. Finally, the coarse graph is recursively refined by reversing the coarsening operations. Then, the ordering of the graph is refined by performing *local permutations*. Each step is shown in Fig. 5.8. Each of these steps will be explained in more detail.

Step 1. Coarsening the graph.   Since the ordering of the graph cannot be computed directly by exhaustively permuting the nodes of the graph, the graph is coarsened enough that one can easily compute the ordering. There can be many different way of coarsening the graph. One criterion that should be considered is that

Figure 5.7: **Isosurface model:** This figure illustrates an isosurface (100M triangles) extracted from a 3D simulation of turbulent fluids mixing. The cache-coherent layout reduces the vertex cache misses by more than a factor of four during view-dependent rendering. As a result, The cache-coherent layouts improve the frame rate by 4 times as compared to prior approaches. A throughput of 90M tri/s (at 30 fps) was achieved on a PC with an NVIDIA GeForce 6800 GPU [YLPM05].

the structure of the graph should be well preserved in the coarsening steps. Otherwise, the computed layout on the coarsened graph may not be a good candidate for the input graph.

To meet this criterion, the graph is partitioned into k chunks (e.g., $k = 4$) by using a graph partitioning library called Metis [KK98]. During the graph partitioning, the number of nodes assigned to each chunk is attempted to be balanced. The number of crossing arcs whose two nodes span two different chunks is attempted to be minimized. Since the crossing arcs are likely to have high arc lengths, it is desirable to reduce the number of such arcs. If the graph is partitioned to k different chunks, the k chunks correspond the nodes of the coarsest graph. In addition, the number of crossing arcs between different chunks is computed to evaluate the cache-oblivious metric. A graph contained in each chunks is recursively partitioned into another k chunks. The original chunk is the coarsened representation of the k different chunks. This process continues until each chunk has less than or equal to k nodes.

Step 2. Ordering the coarsest graph.   Given the coarsest graph consisting of only k different nodes, all possible orderings of its nodes are listed and the costs are computed based on the cache-oblivious metric. A node ordering that has the minimum cost is chosen among all the orderings. Since the coarsest graph contains k nodes, where k is typically four, the number of all the possible ordering is $4!(= 24)$. Therefore, this step takes only minor portion of total processing time of the layout computation method.

Step 3. Refining and local optimization.   Once an ordering of the coarsest graph is computed, the coarsest graph is refined by reversing the coarsening the graph. Also, the node ordering of the coarsest graph is refined to one of the refined graph. Note that by the nature of the coarsening operation, only a node of the graph is expanded to k nodes and other nodes are not changed. Therefore, only one node in the ordering of

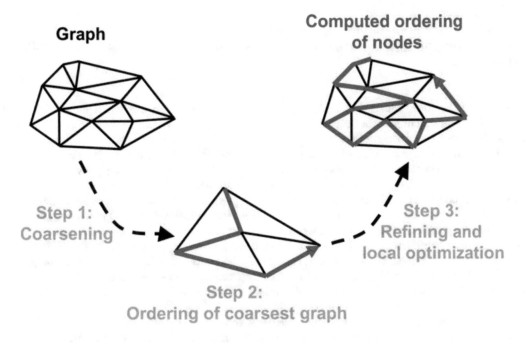

Figure 5.8: **Multi-level layout optimization:** This figure shows three majors steps—coarsening, ordering of the coarsest graph, and refinement and local optimization—of the multi-level layout optimization method.

the graph is expanded to k nodes. Instead of computing a new ordering of the refined graph from scratch, a new ordering of the newly expanded k nodes is computed while keeping the node ordering of other nodes in the refined graph. In order to compute the new ordering of the newly expanded nodes, all the possible local permutations of the newly expanded nodes are generated and an ordering that has minimum cost of the metric is chosen.

Local permutation and metric evaluation.   Local permutations of nodes is performed during Steps 2 and 3. A local permutation affects only a small number of nodes in the layout and changes the arc lengths of those arcs that are incident to these nodes. Therefore, the cost associated with the metric can be efficiently recomputed. Each local permutation involves $k!$ possible orderings for k node. For efficiency, each coarsening operation is restricted to merge $k = 4$ nodes at a time. The number of nodes in the coarsest graph is also limited to 4.

To evaluate a local ordering during the multi-level layout optimization method, a way of computing arc length for a layout of a coarsened graph is necessary. Suppose that an input graph has 80 nodes and the graph is coarsened into a graph consisting of 4 chunks; therefore, 20 nodes are coarsened into each chunk. Then, suppose that an order of those chunks is computed. Then, when an arc consisting of two chunks is computed, the node representing the chunk is treated to have a position in the middle of nodes contained in each chunk. Therefore, for an arc consisting of the first and second chunks in the layout, the arc has arc length 20 computed from 30 minus 10, where 30 and 10 are the middle positions of the first and second chunk.

5.4.2   Analysis

The multi-level construction method explained in the previous section is quite effective and efficient in terms of computing the layout of a graph. This section will discuss optimality and relationship with space filling curves.

Optimality.   It has been found that the multi-level layout optimization method is quite effective for computing a layout that minimizes the cache-oblivious metric. The main reason is that the metric measures the probability that an arc becomes a straddling arc in geometrically increasing block sizes. The chunk sizes during the multi-level layout optimization method vary geometrically. Therefore, it is implicitly considered that chunk sizes likely correspond to block sizes that are geometrically increasing, e.g., power-of-two sized blocks. A graph partitioning method is used to compute coarsened graphs. Note that partitioning a mesh consisting of n nodes into k sets corresponds to computing a cache-aware layout for a block size $\frac{n}{k}$.

In order to see how close the layout computed from the multi-level layout optimization method is to the optimal layout, an optimal layout on a 4 × 4 grid is exhaustively searched given the cache-oblivious metric. The $\beta\Omega$ space-filling curve [Wie02] is the optimal layout closely followed by the Hilbert curve, which confirms conventional wisdom. The $\beta\Omega$ and the Hilbert curves are shown in Fig. 5.9. Compared to the quality of the $\beta\Omega$ space-filling curve, the layout computed from the multi-level construction method shows about 50% higher value for the cache-oblivious metric and shows 60% more cache misses than space filling curves in one of the tested benchmark applications.

There are two main reasons that the multi-level construction method shows lower performance on the uniform grid than the space-filling curve. First, the multi-level layout optimization method does employ local greedy methods. Second, the layout method cannot achieve the optimal partitioning results because partitioning and minimizing the straddling arcs is an NP-hard problem.

Relationship with space-filling curves.   The multi-level layout optimization method shares many similar features with well-known space filling curves like Z-curves, which is constructed in uniform grids. Both the multi-level layout optimization method and typical space-filling curve construction methods construct layouts in a multi-level manner. A multi-level approach constructs a layout, refines it, and extends it locally. The multi-level construction method coarsens the graph by performing graph partitioning, while space-filling curves achieve the optimal partitioning at each recursively level by geometrically partitioning nodes into uniform grids.

The major difference between the multi-level layout optimization method over the common space filling curves is that space-filling curves are typically defined on uniform grids. In contrast, the multi-level layout method can compute cache-coherent layouts for general graphs. In a way, the multi-level layout method is a generalized space-filling curve construction method for general graphs.

5.4.3   Out-of-Core Multi-Level Optimization

One may want to compute cache-coherent layouts of massive meshes. In order to handle massive models, an out-of-core multi-level layout construction method must be designed. At a high level, a set of spatially coherent clusters for an input graph is computed. Each cluster is constructed so that it can be loaded and processed in main memory; typically each cluster has 4K nodes, vertices, or triangles. In order to compute a cache-coherent layout of a massive model, it is critical to construct spatially coherent clusters. Computing such clusters from massive models is a challenging problem. Fortunately, there is an existing technique [YSGM04, YSGM05].

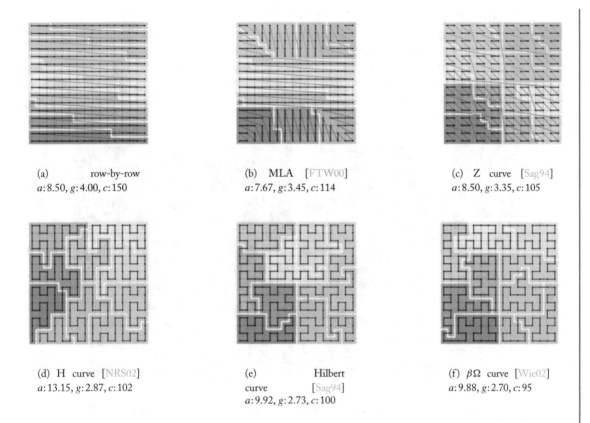

(a)   row-by-row
a: 8.50, g: 4.00, c: 150

(b)   MLA   [FTW00]
a: 7.67, g: 3.45, c: 114

(c)   Z   curve   [Sag94]
a: 8.50, g: 3.35, c: 105

(d)   H   curve   [NRS02]
a: 13.15, g: 2.87, c: 102

(e)   Hilbert
curve   [Sag94]
a: 9.92, g: 2.73, c: 100

(f) $\beta\Omega$ curve   [Wie02]
a: 9.88, g: 2.70, c: 95

Figure 5.9: **Series Layouts and Coherence Measures for a 16 × 16 Grid:** a and g correspond to the arithmetic and geometric mean index difference of adjacent vertices; c denotes the *cut*, or number of edges that straddle cache blocks. Each block except the last contains 27 vertices. MLA is known to minimize a, and $\beta\Omega$ is near-optimal with respect to g for grids. The new cache-oblivious measure, g, correlates better than a with the cut and, hence, the number of cache misses [YL06]. (©IEEE, 2006).

Once spatially coherent clusters are constructed, a layout of clusters is computed based on the cache-oblivious metric. A cache-coherent layout for each cluster is then constructed by sequentially traversing clusters in the cluster layout. To create a cache-coherent layout of a whole input graph, layouts of clusters are simply concatenated in the order of clusters as they appear in the cluster layout.

Processing time. The multi-level layout optimization method has been implemented on a 2.4GHz Pentium-4 PC with 1GB of RAM. The METIS graph partitioning library [KK98] was used for coarsening operations to lay out nodes of an input graph. Source code of the multi-level layout construction method is available at http://gamma.cs.unc.edu/COL/OpenCCL/. The current unoptimized implementation of the out-of-core layout computation can process about 30K triangles per sec. In the case of the St. Matthew model consisting of 372 million triangles, the layout computation takes about 2.6 hours.

5.5   CACHE-COHERENT LAYOUTS OF HIERARCHIES

In this section, a specialized layout technique for hierarchies, especially, bounding volume hierarchies, is reviewed. Bounding volume hierarchies (BVHs) are widely used to accelerate the performance of geometric processing and interactive graphics applications. The applications include ray tracing, visibility culling, collision detection, and geometric computations on large datasets. Most of these algorithms precompute a BVH and traverse the hierarchy at runtime to perform intersection tests or culling. Please refer to Sec. 2.2 for BVH construction methods.

Figure 5.10: **Hugo and 1M Power Plant Models:** The Hugo robot model is placed in the top left of the power plant model, whose overall shape is shown on the right. The performance of collision detection improved by 35–2,600% by using cache-efficient layouts versus other layouts [YM06]. (©Blackwell, 2006).

The leaf nodes of a BVH correspond to the triangles of the original model. The intermediate nodes are the bounding volumes (BVs), commonly represented as spheres, axis-aligned bounding boxes (AABBs), oriented-bounding boxes (OBBs), or convex polytopes. The memory requirements of BVHs can be high for large datasets. For example, the storage cost of a hierarchy of OBBs (an OBB-tree) is approximately 64 bytes per node. As a result, BVHs of large datasets composed of tens of millions of triangles can require gigabytes of space.

The goal is to compute cache-efficient layouts of BVHs to reduce the number of cache misses and improve the performance of BVH-based algorithms. To meet the goal, data layout optimization techniques can be used to place the nodes of a BVH in the memory and reduce the number of cache misses at runtime.

In general, the cache-coherent data optimization method explained in Sec. 5.3 can be used to compute layouts of BVHs. Further performance improvements can be attained by considering two other factors specifically tuned to BVHs:

- **Data access pattern of BVHs.** Before constructing cache-coherent layouts of bounding volume nodes of BVHs, a data access graph representing the access patterns of BVH-based algorithms on BVHs is built. For the mesh case, the basic assumption is that applications access vertices dynamically by traversing the edges of the mesh in a random order. However, BVHs are typically traversed from the root node to leaf nodes. Traversal can proceed along either right or left nodes. Therefore, layout optimization can be improved by accommodating root-to-leaf data access.

- **Specialized construction.** The data layout optimization method explained in Sec. 5.4 assumes that each node and each arc in the data access graph is equally likely to be accessed at runtime. However, this is not true for BVHs. For example, if a bounding volume of a node is comparatively large, it is likely that the bounding volume will be accessed more frequently than smaller volumes during traversal. Therefore, the size of bounding volume is also useful information when developing a BVH layout algorithm.

In this section, a cache-coherent layout algorithm of BVHs that incorporates both optimization methods is developed.

Figure 5.11: **Ray Tracing the Lucy model:** A standard kd-tree based ray tracing algorithm was applied to the Lucy model. A reflective plane is placed behind the Lucy model and the scene also has shadows. A cache-efficient layout of the kd-tree of the Lucy model was computed. The layout improves the performance of ray tracing by up to two times over previous layouts, without any change to the underlying algorithm [YM06]. (©Blackwell, 2006).

5.5.1   Overview of BVH Layout Computation
In this section we define notations related to layouts and give an overview of BVH layout computation.

Notations.   Define n_i^1 as the ith BV node at the leaf level of the hierarchy and n_i^k as a BV node at the kth level of the hierarchy. Also, define $\text{Left}(n_i^k)$ and $\text{Right}(n_i^k)$ to be the left and right child nodes of the n_i^k. A parent node and a grandparent node of the n_i^k are denoted by using $\text{Parent}(n_i^k)$ and $\text{Grand}(n_i^k)$. Formally speaking, a BVH is a directed acyclic graph, $G(N, A)$, where N is a set of BV nodes, n_i^k, and A is a set of directed edges from a node, n_i^k, to each child node, Left(n_i^k) and Right(n_i^k), in the BVH. A layout of a BVH is composed of two parts: a BV layout and a triangle layout. A BV layout of a BVH, $G(N, A)$, is a one-to-one mapping of BVs to positions in the layout, $\varphi : N \rightarrow \{1, \dots, |N|\}$. The goal is to compute a mapping, φ, that minimizes the number of cache misses and the size of working set during the traversal of the BVH at runtime. In addition, a triangle layout is computed to minimize both cache misses and the working set size during BVH traversals.

Data access graph and two localities.   In order to compute a cache-coherent layout of BV nodes of a BVH, the access patterns for the BVH must be analyzed. A layout that minimizes the number of cache misses to the BVH during collision queries must be computed. Collision queries continue to traverse BVHs as long as each query between two BVs reports a collision between them. The goal is to minimize the number of cache misses and the size of working set during the traversal. There are two different localities, parent-child locality and spatial locality, which arise during the traversal.

1. **Parent-child locality.** Once a node of a hierarchy is accessed by a search query, it is likely that its child nodes will be accessed quickly. For example, in frame i of Fig. 5.12, if the root node of the BVH is accessed, its two child nodes, n_1^3 and n_5^3, are also likely to be accessed quickly. Moreover, after n_1^3 is accessed during frame i, its child nodes are likely to be accessed in the next frame.

2. **Spatial locality.** Whenever a node is accessed by a search query, other nodes in close proximity are also highly likely to be accessed by other search queries. For example, collisions or contacts between two objects occur in small localized regions of a mesh. Therefore, if a node of a BVH is accessed, other nearby nodes are either colliding or are in close proximity and may be accessed soon. In frame $i + 1$ of Fig. 5.12, if one of two nodes, n_4^1 and n_7^1, is accessed, the other node is also likely to be accessed during that frame or subsequent frames.

The layout construction method described next accommodates these two localities and uses them to compute the layout of a BVH.

Probability computations.   In addition to the different localities, the strength of the locality between two nodes in the BVH can also be computed. For example, suppose that a left node has much bigger bounding volume than a right node has. In this case, it is more likely that the left node will be accessed than the right node. In other words, the parent-child locality between the parent node and the left node is stronger than one between the parent node and the right node. In order to quantify strength of the parent-child locality, the probability of that a BV of a second object has degenerated to a point is computed. This assumption simplifies the probabilistic model and has a strong correlation even when the BV has other shapes [YM06]. Given this assumption, a probability, $Pr(n)$, that a node n will be accessed when its parent node Parent(n) is accessed is defined as the following:

$$Pr(n) = \frac{\text{Vol}(BV(n) \cap BV(\text{Parent}(n)))}{\text{Vol}(BV(\text{Parent}(n)))}, \tag{5.6}$$

where Vol($BV(n)$) means the volume of a BV node $BV(n)$.

5.5.2   Layout Optimization

The next section describes an algorithm to compute cache-coherent layouts of BVHs. Knowing the cache parameters and the block size is needed to compute how many BV nodes fit into a given cache block. It is

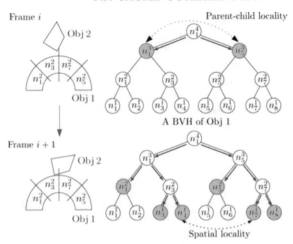

Figure 5.12: Two localities within BVHs: Two successive frames from a dynamic simulation and the change in access patterns (shown with blue arrows) of a BVH. In this simulation, object 2 drops on object 1, as shown on the left. The access pattern of the BVH of object 1 during each frame is shown on the right. The BVs from the 2nd level in the BVH are shown within object 1 on the left. The figure also illustrates the front traversed within each BVH during each frame in green. The top BVH shows the parent-child locality, when the root node, n_1^4, of the BVH of object 1 collides with the BVs of objects 2. During frame $i + 1$, object 2 is colliding with object 1. In this configuration, the BVs n_3^2 and n_7^2 (and their sub-nodes) are accessed due to their close spatial locality [YM06]. (©Blackwell, 2006).

possible to decompose the BVH into a set of clusters using this knowledge. For example, the size of each cluster may be equal to the size of the cache block. This algorithm does not assume any particular cache size and constructs a layout that works well with any cache parameter. In order to achieve this, the layout computations are performed recursively. At each recursion level, the layout construction method has two main steps: (1) clustering and (2) ordering clusters. The BVH is first decomposed into a set of clusters. The second step computes an ordering of clusters. Therefore, this allows a cache-efficient ordering of the clusters to be computed at each level of recursion.

Cluster computation. The goal for computing clusters is to store the BV nodes that are accessed together because of the parent-child locality into the same cluster. This minimizes the number of cache misses. In order to achieve this goal, the probabilities for each parent-child locality are computed according to the probabilistic model shown in Eq. (5.6). Nodes are then assigned by traversing the BVH from the root into a cluster. During the traversal, a front of nodes is maintained and the node that has the highest probability of belonging to the cluster is greedily assigned. Although, this greedy cluster computation method may not maximize the sum of the probabilities of nodes contained in the cluster, it works well in most cases.

Layouts of clusters. Given the computed clusters at each level of recursion, a cache-oblivious ordering of the clusters is computed by considering their spatial locality. The root cluster is placed at the beginning of the ordering of clusters because traversal typically starts at the root node of the BVH. In order to compute an ordering of child clusters, a graph is constructed with the child clusters as the nodes of the graph. An arc is then computed between two clusters if they are in close proximity, i.e., if their BVs overlap. Then, the probability that a BV of a cluster has collided given that a BV of another cluster has collided is computed based on the probability formulation described in Eq. (5.6). An example of an undirected graph between two child clusters is shown in the middle BVH of Fig. 5.13. Once the graph is computed, a cache-oblivious layout is computed from the graph that represents the access patterns between the child clusters. This is performed using the cache-oblivious mesh layout algorithm explained in the previous section. An example of a cache-oblivious layout of a complete tree is shown in the rightmost figure of Fig. 5.13.

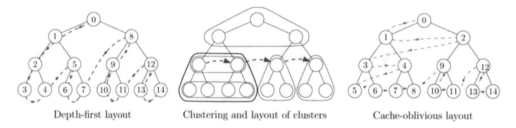

Figure 5.13: **Layout computation of a BVH:** A depth-first layout of a BVH is shown in the leftmost figure and a cache-oblivious layout of the same tree is shown in the rightmost figure. The number within each BV node in the leftmost and the rightmost figures is an index of the ordering of BVs in the layout. The middle figure shows the output of the clustering step. The topmost cluster is the root cluster and the rest are child clusters. Directed edges (shown in blue) indicate ordering between clusters. The middle figure shows that leftmost cluster is merged with its neighboring cluster [YM06]. (©Blackwell, 2006).

5.6   APPLICATIONS

In this chapter, cache-coherent layout techniques and how they can be applied to rasterization, ray tracing, iso-surface extraction, and collision detection are discussed. The layout techniques are applied to: (1) triangle meshes; (2) view-dependent meshes; and (3) bounding volume hierarchies.

5.6.1   Triangle Meshes

Triangle meshes are one of most widely used data representations in computer graphics. In order to compute a cache-coherent layout for a mesh, the mesh connectivity are used to compute a data access graph for the layout. The layouts are applied to the problem of computing iso-contouring. Iso-contouring is widely used in geographic information systems (GIS). It traverses an input triangular mesh and extracts a list of triangles that has an iso-value specified by a user.

The algorithm used is based on seed sets [vKvOB+97] to extract the iso-contour. Seed sets provide the initial access points to extract iso-contours. When using seed sets, the running time of the iso-contouring is dominated by the traversal of the triangles having the specified iso-value.

In order to compare the performance of different storage layouts, the iso-surface are computed using different layouts, including cache-coherent layouts, spectral layouts [IL05], and layouts geometrically sorted along X/Y/Z directions (vertices sorted by their positions along the chosen direction) for a Puget sound terrain model (see Fig. 5.14) consisting of 134 million triangles.

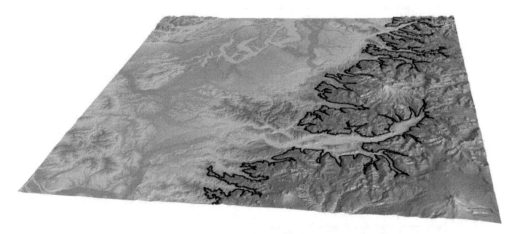

Figure 5.14: **Puget Sound contour line:** An iso-contour (shown in black) extracted from an unstructured terrain model of the Puget Sound consisting of 143M is shown. The largest component (223K edges) of the level set have been extracted at 500 meters of elevation. Cache-oblivious layouts improve the performance of iso-contour extraction algorithm by more than an order of magnitude [YLPM05].

The iso-contour algorithm extract contours that are parallel to XY-plane, i.e., the extracted iso-contour has same Z-value (or height). In addition, a ridge line is extracted by traversing the terrain model upward from a seed triangle that is a saddle. To examine the performance, the same iso-contour and ridge are extracted twice. During the second extraction process, all the necessary data has been loaded into main memory. The cost of the cache coherent layout is represented by the difference between the first and second extraction steps. The major computational bottleneck during the second extraction is the data access time between L1/L2 cache and main memory, while the bottleneck is the data access time between disk and main memory during the first extraction.

Iso-contouring performance with different layouts is shown at Tab. 5.1.

Analysis.   When an iso-contour is extracted along the Z axis, the best performance is achieved when the Z-axis is sorted. The sorted Z-axis layout is also effective when processing such iso-contour queries, whose access pattern matches well to the Z-axis layout. Even in this case, the cache-oblivious mesh layout shows performance that is close to the best performance achieved by the Z-axis sorted layout.

When a ridge is extracted, the best performance is achieved by X-sorted layout. In this case, the cache-oblivious layout shows high performance close to that of X-axis sorted layout. However, the performance with other layouts shows lower performance compared to the cache-oblivious mesh layout. Since the cache-oblivious mesh layout is not optimized for a specific runtime query and optimized instead for various runtime queries, high performance can be achieved for multiple runtime queries. Similar results are obtained when the same iso-contour and the same ridge is extracted for the second time. These tests demonstrate the nature of the cache-oblivious layouts of the mesh, since the performance improvement can be observed when the bottleneck is between disk and main memory and between L1/L2 cache and main memory.

The cache-coherent layout construction methods can also be applied to iso-surface extraction on tetrahedron meshes. More detail information can be found at Yoon et al. [YL06].

Table 5.1: **Iso-contouring:** Time in seconds (on a 1.3 GHz Linux PC with 2 GB of memory) for extracting an iso-contour (or equivalent geometric queries) for Puget sound terrain model stored each in five different mesh layouts: cache-oblivious, with vertices sorted by X/Y/Z geometric coordinate, and spectral sequencing. The time for second immediate re-computation of the same contour when all the cache levels in the memory hierarchy have been loaded are listed in parentheses. In all the cases, the performance of the cache-oblivious layout is comparable to the one optimized for the particular geometric query. This demonstrates the benefit of the layout for general applications. The table is modified from Tab. 4 shown in [YLPM05].

Model	Puget Sound	
Out. edg.	223K (Contour)	14K (Ridge)
Cac. Obl.	026 (000.5)	003 (000.03)
Geom. X	232 (227.8)	001 (000.04)
Geom. Y	218 (215.5)	195 (185.10)
Geom. Z	011 (000.6)	135 (113.81)
Spec. Seq.	150 (127.3)	023 (000.04)

5.6.2   View-Dependent Meshes

View-dependent meshes are frequently used to improve the performance of rendering massive models consisting of hundreds of millions of triangles. View-dependent meshes can provide smooth varying multi-resolution, which reduces visual artifacts like popping. Therefore, view-dependent meshes are typically used for high-quality view-dependent rendering of massive models. However, most of existing techniques of rendering view-dependent meshes have achieved very low rendering throughput since it is expensive to compute cache-coherent triangle strips for view-dependent meshes because view dependent meshes must change their mesh connectivity every frame.

The preferred method is a clustered hierarchy of progressive meshes (CHPM), a representation proposed in Quick-VDR system [YSGM04]. The CHPM representation has been shown to provide interactive view-dependent rendering performance for massive models composed of hundreds of millions of triangles. CHPM representations contain multiple progressive meshes. Each progressive mesh is stored with its base mesh and an array of vertex splits to refine the base mesh. At runtime, a front that contains vertices representing the current multi-resolution level is maintained. The front is updated based on an LOD metric. In addition, a second front that contains triangles representing the current resolution of the mesh is maintained. The triangle front associated with each progressive mesh is initialized with triangles stored in the base mesh.

Cache-oblivious layouts for base meshes associated with progressive meshes are then computed. When a vertex v_a is split into two vertices at runtime by performing a refinement operation, two new triangles are added to the triangle front and a triangle associated with the vertex v_a is removed. In order to maintain the cache-coherent layout for view-dependent mesh, the two new triangles occupy the same position as the deleted triangle.

The rendering performance of CHPM representations have been compared with different layouts, including the cache-oblivious layout, universal rendering sequences [BG02], Hoppe's rendering sequence [Hop99a], and a Z-curve [Sag94]. The comparison is based on measuring the average cache miss ratio (ACMR), which is defined as the ratio of the number of vertex cache misses to the number of rendered triangles given a chosen vertex cache size. To compute the ACMR with different layouts, a simulated GPU vertex cache with FIFO replacement was developed as proposed in [Hop99a].

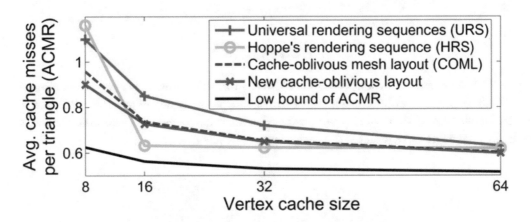

Figure 5.15: **Comparison with Other Rendering Sequences in Bunny Model:** ACMRs of cache-oblivious layout (**COL**) are close to optimal ACMRs. Also, COL consistently outperforms the universal rendering sequence (**URS**), cache-oblivious mesh layout (COML), and Hoppe's rendering sequence (**HRS**) at cache size 8 and 64; HRS is optimized at cache size 12 or 16.

Figure 5.15 shows ACMRs of different rendering sequences on the Stanford bunny model. The cache-oblivious layout of the bunny model labeled 'new cache-oblivious layout' shows the lowest ACMR values compared to the universal rendering sequence of all other tested GPU vertex cache sizes. Compared to Hoppe's rendering sequence, which is optimized around a GPU vertex cache size 16, the cache-oblivious layout shows worse performance at that specific cache size. However, the cache-oblivious layout shows better performance at other tested vertex cache sizes of 8 and 64. This result demonstrates the nature of cache-oblivious layout, which is optimized to handle various cache sizes. Note that ACMR values of the cache-oblivious layout are very close to optimal ACMRs, which is proposed at [BG02].

ACMRs of different rendering layouts have been computed for a power plant model (see Fig. 5.2). The results for the power plant are similar to those measured for the Stanford bunny. When comparing power plant ACMRs to those of the Z-curve method (space-filling curves computed on uniform grids), the cache-oblivious layout shows a much higher cache hit ratio. Since the power plant model has very irregular distribution of geometry, the quality of space-filling curves for power plant may not be as high as a cache-oblivious mesh layout method. This result supports the claim that a cache oblivious layout technique is a generalized layout construction method for unstructured meshes, while classical space-filling curves like Z-curves works best on uniform grids.

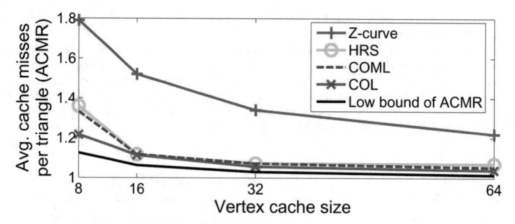

Figure 5.16: **Comparison with Space-Filling Curve in Power Plant:** Our new cache-oblivious layout (COL) consistently performs better than Z-curve, Hoppe's rendering sequences (HRS), and cache-oblivious mesh layout (COML) on a power plant model, which has irregular geometric distribution. These results are also congruent to what the cache-oblivious metric predicts.

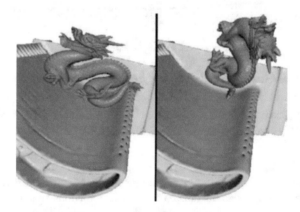

Figure 5.17: **Dynamic Simulation between Dragon and Turbine Models:** This image sequence shows discrete positions from a dynamic simulation between dragon and CAD turbine models. A 38–215% performance improvement can be achieved in collision detection by using cache-efficient layouts of the OBB-tree over other tested layouts [YM06]. (©Blackwell, 2006).

5.6.3   Bounding Volume Hierarchies

Bounding volume hierarchies are widely used to perform proximity queries such as collision detection, visibility queries, minimum separation distance, and ray tracing. To perform proximity queries, a runtime algorithm typically traverses the BVH from the root node to one of the leaf nodes. During traversal, the major performance bottleneck occurs during data access. In order to improve the performance of collision detection and ray tracing, cache-coherent layouts are computed as BVHs as explained in Sec. 5.5.

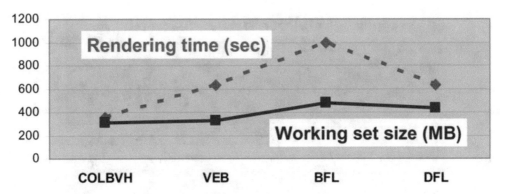

Figure 5.18: Performance of Ray Tracing: Average render time and the size of working set during ray tracing of the Lucy model with 28 million triangles are shown with different layouts. By using the cache-oblivious layout, a 77–180% performance improvement can be achieved in of ray tracing and reduce the working set size by 7–55% [YM06]. (©Blackwell, 2006).

Collision detection.   Performance testing of collision detections is based on three different benchmarks. The first benchmark detected collisions between the Stanford bunny model and the dragon model. In this benchmark, the bunny model is dropped on top of the dragon model and a rigid body simulation between two models is performed. The second detected collisions between the dragon model and CAD turbine model (see Fig. 5.17). The third placed a robot model in the top left side of a power plant model (see Fig. 5.10). The set of benchmarks is labeled Benchmark 3-a. The last configuration placed the robot model in the furnace room, which is at the middle of the power plant model. The last benchmark is labeled Benchmark 3-b.

Figure 5.19 shows the processing time and working set size during performing collision checking and hierarchy traversal with different layouts. Cache-oblivious layouts of BVHs (COLBVH) shows higher performance than other tested layouts including van Emde Boas layout (VEB) [vEB77], depth-first layout (DFL), breadth-first layout (BFL), and cache-oblivious mesh layout technique explained in Sec. 5.3. Moreover, the cache-oblivious layouts show comparable performance to the cache-aware layouts of BVHs (CALBVH) as shown in Benchmark 1 and 2. Since the main computational bottleneck of performing collision detection is at reading data from disk, set disk, and cache block sizes are set to 4KB for cache-aware layout computation.

Ray tracing.   A kd-tree for static models has been considered as one of best acceleration hierarchies for ray tracing implementation [Wal04]. To test various layouts of kd-trees, the representation of intermediate kd-nodes proposed in Wald [Wal04] have been modified to have the left and right child indices. This implementation causes the size of each kd-node becomes 16 bytes as opposed to 8 bytes commonly used in the state-of-the-art kd-tree representation.

Different cache layouts of a Lucy model (see Fig. 5.11) consisting of 28 million triangles have been compared. The layouts tested include the cache-oblivious layouts of BVHs, van Emde Boas layouts, depth-first, and breadth-first layouts. Figure 5.11 shows the ray tracing time and working set size when using different layouts. Performance improved by 77–180% when using the cache-oblivious layouts over other layout methods. This performance improvement is mainly achieved by reducing working set size.

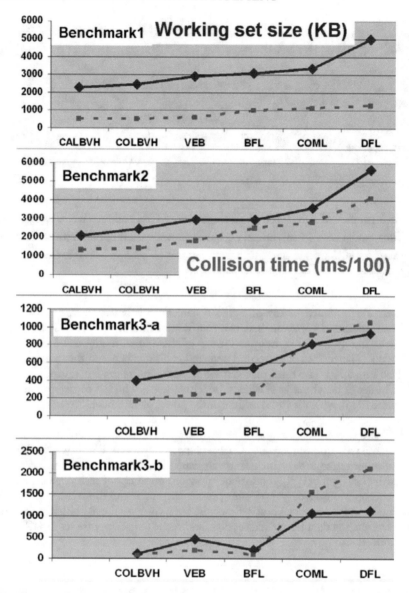

Figure 5.19: Performance of Collision Detection: Average collision query time and the size of working set for collision detection. Note the performance of layouts like VEB, DFL and BFL and compare them with cache oblivious layouts layouts like COLBVH and CALBVH. VEB is the van Emde Boas layout, DFL and BFL are the depth-first and breadth-first layouts, respectively. Overall, performance improved 26–2,600% for collision queries based on reduced working set size and caused fewer cache misses. Moreover, the performance of cache-oblivious layout (COLBVH) is comparable to that of cache-aware layouts (CALBVH) (in the first and second benchmarks) and consistently shows better performance over other layouts [YM06]. (©Blackwell, 2006).

5.7   DISCUSSION

This chapter covered the issues involving data layout techniques, especially for meshes and bounding volume hierarchies. The techniques are applicable to raster-based rendering, ray tracing, collision detection, and other functions which must process huge data volumes at real-time or near real-time rates. The data layout techniques produce cache-coherent layouts of meshes and BVHs that minimize the number of cache misses when accessing mass storage resident data. The dominant concept is to match the data layout closely to the data access pattern of runtime applications.

There are many other approaches that deserve further investigation. Current optimization methods are compute-intensive and consume large amounts of processing time to compute layouts for hundreds of millions of triangles and triangle meshes. Designing a faster layout construction method and still being able to produce high-quality layouts will further expand the usability of layout optimization methods.

The chapter examined construction techniques for cache-coherent layouts without modifying the algorithms. Research is needed to examine the benefits of re-ordering the underlying order of the algorithms.

The third research area extends the generalization of the layout algorithms. Even though performance has improved in many different applications, in most of cases, the increase resulted from improved disk access time. It requires further investigation to improve the performance of applications whose major bottleneck is data access between L1/L2 caches and main memory.

5.8   FURTHER READING

Graph theory has been a rich source of research applicable to layout optimization methods. Recent results on layout methods with various metrics are available in extensive survey done by Diaz et al. [DPS02]. Out-of-core issues for visualization problems are covered extensively in a survey by Silva et al. [SCC$^+$02]. Although, not discussed here, designing cache-coherent algorithms may be a good alternative to designing cache-coherent layout methods. Cache-oblivious methods are well explained in the survey done by Arge et al. [ABF04]. Vitter [Vit01] wrote an excellent survey on external memory algorithms and data structures.

CHAPTER 6

Conclusions

Massive model visualization remains an active area of research in computer graphics, scientific visualization, GIS, urban planning, and other areas. The recent advances in acquisition technologies and development of consumer applications such as Google Earth, the Visible Human Project, and Microsoft Virtual Earth show a mass market potential of the underlying ideas. In this monograph, we have given an overview of multiple approaches to accelerate massive model rendering. We focused on algorithms for visibility culling, complexity reduction, and memory layout and cache-friendly techniques. Although, most of our focus is on improving the performance of rasterization algorithms, a number of the same methods and representations are also applicable to ray tracing.

There are many aspects of data representations, model manipulation and rendering, and system design that we introduced in Ch. 1 and have not fully covered here.

Dynamic models. The techniques we covered are effective mainly for static models. Additional work must be done to address dynamic (i.e., where parts of the model move with respect to one another or change during a session) models. Example uses include time-dependent simulations of scientific phenomena, 4D real-world events that capture the motion of the objects in a scene, and physically based simulations. Most visibility and complexity reduction techniques use a hierarchical representation of the model to accelerate the computation. As the model changes, the hierarchical representation must be updated and/or recomputed. Recomputing the structures at 10Hz or faster to keep rendering rates acceptable must be accomplished while basic rendering is occurring. For example, in the context of bounding volume hierarchies (BVH) based ray tracing, there are three possible options for updating the BVH: (1) incremental BV re-fitting by traversing the BVH; (2) re-construction of the BVH from the scratch; and (3) selective restructuring method [YCM07]. Each of these methods has its own advantages and disadvantages. Further research is required for other types of applications, hierarchies, and model simplification techniques.

Integration with interactive techniques. Once we render massive models interactively, users will like to interact with those models in a natural manner. These include development of novel 2D or 3D interaction techniques that can handle the complexity of these models. Collision detection and interference checking add substantial value to basic visualization to help determine the quality of the model or to support haptics. Many current collision detection algorithms rely on acceleration hierarchies (e.g., bounding volume hierarchies, kd-trees) that may be able to be shared with the rendering hierarchy. Currently, most collision detection methods employ their own acceleration hierarchies. Ideally, we would like to develop a unified representation that can be used both for collision checking and visualization. Some early work is the dual hierarchy based on simplified models used for collision detection and view-dependent rendering [YSLM04].

Global illumination and photo-realistic rendering. Most of techniques explained in this monograph mainly use local illumination models, which are simplified lighting methods that can utilize the capabilities of current GPUs. As we get higher rendering performance, it is likely that global illumination methods will be more widely used to achieve more photo-realistic effects. However, global illumination methods inherently require much more computation time than local illumination methods. We expect that significant research efforts will be required to make global illumination methods more widely accepted for massive model visualization.

The advances made in massive model visualization over the past 5 years have already exceeded those of the previous 45 years. The techniques we describe here are essential to making massive model visualization a commodity that is widely used in the next 10 years. Other issues affecting visualization of complexity will continue to appear, so the field is ripe for continued research as long as users generate more data than commodity hardware can handle.

Bibliography

[ABCO+01]   M. Alexa, J. Behr, D. Cohen-Or, S. Fleishman, D. Levin, and C.T. Silva. Point Set Surfaces. In *Proceedings of IEEE Visualization 2001*, pp. 21–28, 2001. DOI: 10.1109/VISUAL.2001.964489

[ABD+03]   S. Alstrup, M.A. Bende, E.D. Demaine, M. Farach-Colton, T. Rauhe, and M. Thorup. Efficient tree layout in a multilevel memory hierarchy. *Computing Research Repository (CoRR)*, 2003.

[ABF04]   L. Arge, G. Brodal, and R. Fagerberg. Cache oblivious data structures. *Handbook on Data Structures and Applications*, ACM, New York, NY, 2004.

[ACW+99]   D. Aliaga, J. Cohen, A. Wilson, H. Zhang, C. Erikson, K. Hoff, T. Hudson, W. Stuerzlinger, E. Baker, R. Bastos, M. Whitton, F. Brooks, and D. Manocha. MMR: An integrated massive model rendering system using geometric and image-based acceleration. In *Proceedings of ACM Symposium on Interactive 3D Graphics*, pp. 199–206, 1999. DOI: 10.1145/300523.300554

[AL97]   D.G. Aliaga and A.A. Lastra. Architectural Walkthroughs Using Portal Textures. In *Proceedings of IEEE Visualization 1997*, pp. 355–362, 1997. DOI: 10.1109/VISUAL.1997.663903

[AM96]   R. Abarbanel and W. McNeely. Flythru the boeing 777. *ACM SIGGRAPH Visual Proceeding*, 1996. DOI: 10.1145/253607.253800

[AM00]   U. Assarsson and T. Möller. Optimized view frustum culling algorithms for bounding boxes. *Journal of Graphic Tools*, 5, 2000.

[Ame07]   The american heritage dictionary of the english language. Houghton Mifflin Company, Boston, MA, 2007.

[AMH02]   T. Akenine-Möller and E. Haines. *Real-Time Rendering, Second Edition*. A K Peters, Wellesley, MA, 2002.

[App68]   A. Appel. Some techniques for shading machine renderings of solids. In *AFIPS 1968 Spring Joint Computer Conf.*, volume 32, pp. 37–45, 1968.

[ARB90]   J.M. Airey, J.H. Rohlf, and F.P. Brooks, Jr. Towards image realism with interactive update rates in complex virtual building environments. *Computer Graphics* 24(2):41–50, March 1990. DOI: 10.1145/91385.91416

[AS94]   P.K. Agarwal and S. Suri. Surface approximation and geometric partitions. In *Proceedings 5th ACM-SIAM Sympos. Discrete Algorithms*, pp. 24–33, 1994. DOI: 10.1137/S0097539794269801

[ASNB00]   C. Andujar, C. Saona, I. Navazo, and P. Brunet. Integrating occlusion culling and levels of detail through hardly-visible sets. *Computer Graphics Forum*, 19(3):499–506, 2000. DOI: 10.1111/1467-8659.00442

[ASVN00]   C. Andújar, C. Saona-Vázquez, and I. Navazo. LOD visibility culling and occluder synthesis. *Computer-Aided Design*, 32(13):773–783, October 2000. DOI: 10.1016/S0010-4485(00)00067-1

[AV88]   A. Aggarwal and J.S. Vitter. The input/output complexity of sorting and related problems. *Communications of the ACM*, 31:1116–1127, 1988. DOI: 10.1145/48529.48535

[BD06]   L. Baboud and X. Décoret.:: Rendering geometry with relief textures. In C. Gutwin and S. Mann, Eds., *Graphics Interface*, pp. 195–201. Canadian Human-Computer Communications Society, 2006.

[BG02]   A. Bogomjakov and C. Gotsman. Universal rendering sequences for transparent vertex caching of progressive meshes. In *Computer Graphics Forum*, pp. 137–148, 2002. DOI: 10.1111/1467-8659.00573

[BGM+07]   L. Borgeat, G. Godin, P. Massicotte, G. Poirier, F. Blais, and J. Beraldin. Analyzing large multi-scale datasets: The case of the virtual mona lisa. *IEEE Computer Graphics and Applications*, 2007. DOI: 10.1109/MCG.2007.162

[BGMP07]   F. Bettio, E. Gobbetti, F. Marton, and G. Pintore. High-quality networked terrain rendering from compressed bitstreams. In *Proceedings ACM Web3D International Symposium*, pp. 37–44. ACM Press, New York, April 2007. DOI: 10.1145/1229390.1229396

[BHS98]   J. Bittner, V. Havran, and P. Slavik. Hierarchical visibility culling with occlusion trees. In *Proceedings of Computer Graphics International '98*, pp. 207–219, 1998. DOI: 10.1109/CGI.1998.694268

[Bit02]   J. Bittner. *Hierarchical Techniques for Visibility Computations*. Ph.D. thesis, Department of Computer Science and Engineering, Faculty of Electrical Engineering, Czech Technical University in Prague, October 2002.

[Bro92]   F. Brooks. Walkthrough project: Final technical report to national science foundation computer and information science and engineering. Technical report, University of North Carolina-Chapel Hill, Computer Science, TR92-026, 1992.

[Bru07]   B. Bruderlin. Interviews3d - a platform for interactive handling of massive data sets. *IEEE Computer Graphics and Applications*, 2007. DOI: 10.1109/MCG.2007.153

[BSGM02]   B. Baxter, A. Sud, N. Govindaraju, and D. Manocha. GigaWalk: Interactive walkthrough of complex 3D environments. *Proceedings of Eurographics Workshop on Rendering*, pp. 203–214, 2002.

[Bux02]   B. Buxton. Less is more (more or less), in the invisible future: The seamless integration of technology into everyday life. McGraw-Hill, 2002.

[BW03]   J. Bittner and P. Wonka. Visibility in computer graphics. *Journal of Environment and Planning B: Planning and Design*, 30(5):729–756, 2003. DOI: 10.1068/b2957

[BWPP04]   J. Bittner, M. Wimmer, H. Piringer, and W. Purgathofer. Coherent hierarchical culling: Hardware occlusion queries made useful. *Computer Graphics Forum*, 23(3):615–624, 2004. DOI: 10.1111/j.1467-8659.2004.00793.x

[BWW01]   J. Bittner, P. Wonka, and M. Wimmer. Visibility preprocessing for urban scenes using line space subdivision. In *PG '01: Proceedings of the 9th Pacific Conference on Computer Graphics and Applications*, p. 276, Washington, DC, IEEE Computer Society, 2001. DOI: 10.1109/PCCGA.2001.962883

[Cat74]   E.E. Catmull. *A Subdivision Algorithm for Computer Display of Curved Surfaces*. Ph.D. thesis, Dept. of CS, University of Utah, December 1974.

[CBWR07]   J.P. Charalambos, J. Bittner, M. Wimmer, and E. Romero. Optimized hlod refinement driven by hardware occlusion queries. In *Advances in Visual Computing (Third International Symposium on Computer Vision – ISVC 2007)*, Lecture Notes in Computer Science, volume 4841, pp. 106–117. Springer, November 2007. DOI: 10.1007/978-3-540-76858-6-11

[CC78]   E. Catmull and J. Clark. Recursively generated B-spline surfaces on arbitrary topological meshes. *Computer-Aided Design*, 10:350–355, September 1978. DOI: 10.1016/0010-4485(78)90110-0

[CCMS97]   A. Ciampalini, P. Cignoni, C. Montani, and R. Scopigno. Multiresolution decimation based on global error. *The Visual Computer*, 13(5):228–246, June 1997. DOI: 10.1007/s003710050101

[CDBG+07] P. Cignoni, M. Di Benedetto, F. Ganovelli, E. Gobbetti, F. Marton, and R. Scopigno. Ray-Casted BlockMaps for Large Urban Models Visualization. In *Computer Graphics Forum (Proceedings of Eurographics)*, 2007. DOI: 10.1111/j.1467-8659.2007.01063.x

[CGG+03a] P. Cignoni, F. Ganovelli, E. Gobbetti, F. Marton, F. Ponchio, and R. Scopigno. BDAM – batched dynamic adaptive meshes for high performance terrain visualization. *Computer Graphics Forum*, 22(3):505–514, Proceedings Eurographics 2003. September 2003. DOI: 10.1111/1467-8659.00698

[CGG+03b] P. Cignoni, F. Ganovelli, E. Gobbetti, F. Marton, F. Ponchio, and R. Scopigno. Planet-sized batched dynamic adaptive meshes (p-bdam). In *Proceedings IEEE Visualization*, pp. 147–155, Seattle, WA, Computer Society Press. October 2003. DOI: 10.1109/VISUAL.2003.1250366

[CGG+04] P. Cignoni, F. Ganovelli, E. Gobbetti, F. Marton, F. Ponchio, and R. Scopigno. Adaptive tetrapuzzles: Efficient out-of-core construction and visualization of gigantic multiresolution polygonal models. *ACM Transactions on Graphics*, 23(3):796–803, August 2004. DOI: 10.1145/1015706.1015802

[CGG+05] P. Cignoni, F. Ganovelli, E. Gobbetti, F. Marton, F. Ponchio, and R. Scopigno. Batched multi triangulation. In *Proceedings IEEE Visualization*, pp. 207–214, Conference held in Minneapolis, MI, IEEE Computer Society Press. October 2005. DOI: 10.1109/VISUAL.2005.1532797

[Che95] S.E. Chen. Quicktime VR - an image-based approach to virtual environment navigation. In R. Cook, Ed., *SIGGRAPH 95 Conference Proceedings*, Annual Conference Series, Los Angeles, pp. 29–38. ACM SIGGRAPH, Addison-Wesley, August 1995. DOI: 10.1145/218380.218395

[CKS03] W.T. Corrêa, J.T. Klosowski, and C.T. Silva. Visibility-based prefetching for interactive out-of-core rendering. In *Proceedings of PVG 2003 (6th IEEE Symposium on Parallel and Large-Data Visualization and Graphics)*, pp. 1–8, 2003. DOI: 10.1109/PVGS.2003.1249035

[Cla76] J.H. Clark. Hierarchical geometric models for visible surface algorithms. *Communications of the ACM*, 19(10):547–554, October 1976. DOI: 10.1145/360349.360354

[CM95] S. Coleman and K. McKinley. Tile size selection using cache organization and data layout. *SIGPLAN Conference on Programming Language Design and Implementation*, pp. 279–290, 1995. DOI: 10.1145/207110.207162

[CMRS03] P. Cignoni, C. Montani, C. Rocchini, and R. Scopigno. External memory management and simplification of huge meshes. In *IEEE Transaction on Visualization and Computer Graphics*, pp. 525–537, 2003. DOI: 10.1109/TVCG.2003.1260746

[COCSD03] D. Cohen-Or, Y.L. Chrysanthou, C.T. Silva, and F. Durand. A survey of visibility for walk-through applications. *IEEE Transactions on Visualization and Computer Graphics*, 9(3):412–431, July/September 2003. DOI: 10.1109/TVCG.2003.1207447

[COFHZ98] D. Cohen-Or, G. Fibich, D. Halperin, and E. Zadicario. Conservative visibility and strong occlusion for viewspace partitioning of densely occluded scenes. *Computer Graphics Forum*, 17(3):243–254, ISSN 1067-7055. 1998. DOI: 10.1111/1467-8659.00271

[Coo84] R.L. Cook. Shade trees. *Computer Graphics (SIGGRAPH 84 Proceedings)*, 18(3):223–231, July 1984. DOI: 10.1145/964965.808602

[COZ98] D. Cohen-Or and E. Zadicario. Visibility streaming for network-based walkthroughs. In *Graphics Interface '98*, pp. 1–7, June 1998.

[CSAD04] D. Cohen-Steiner, P. Alliez, and M. Desbrun. Variational shape approximation. *ACM*

Transactions on Graphics, 23(3):905–914, August 2004. DOI: 10.1145/1015706.1015817

[CT96]   S. Coorg and S. Teller. Temporally coherent conservative visibility. In *Proceedings 12th Annu. ACM Symp. Comp. Geom.*, pp. 78–87, 1996. DOI: 10.1145/237218.237242

[CT97]   S. Coorg and S. Teller. Real-time occlusion culling for models with large occluders. In *1997 Symposium on Interactive 3D Graphics*, pp. 83–90, 1997. DOI: 10.1145/253284.253312

[CVM⁺96]   J. Cohen, A. Varshney, D. Manocha, G. Turk, H. Weber, P. Agarwal, F.P. Brooks, Jr., and W. Wright. Simplification envelopes. In H. Rushmeier, Ed., *SIGGRAPH 96 Conference Proceedings*, Annual Conference Series, pp. 119–128. ACM SIGGRAPH, Addison-Wesley, held in New Orleans, Louisiana, 04-09 August 1996. DOI: 10.1145/237170.237220

[DD02]   F. Duguet and G. Drettakis. Robust epsilon visibility. In J. Hughes, Ed., *Proceedings of ACM SIGGRAPH 2002*. ACM Press / ACM SIGGRAPH, July 2002.

[DDTP00]   F. Durand, G. Drettakis, J. Thollot, and C. Puech. Conservative visibility preprocessing using extended projections. In *SIGGRAPH 00 Conference Proceedings*, pp. 239–248, 2000. DOI: 10.1145/344779.344891

[Dee95]   M.F. Deering. Geometry compression. In *ACM SIGGRAPH*, pp. 13–20, 1995. DOI: 10.1145/218380.218391

[DGBGP05]   P. Diaz-Gutierrez, A. Bhushan, M. Gopi, and R. Pajarola. Constrained strip generation and management for efficient interactive 3d rendering. In *Computer Graphics International*, pp. 115–121, 2005. DOI: 10.1109/CGI.2005.1500388

[dL04]   R. de Toledo and B. Levi. Extending the graphic pipeline with new GPU-accelerated primitives. In *Proceedings 24th gOcad Meeting*, Nancy, France, 2004.

[dLP07]   R. de Toledo, B. Levy, and J.-C. Paul. Iterative methods for visualization of implicit surfaces on gpu. In *ISVC, International Symposium on Visual Computing*, Lecture Notes in Computer Science, Lake Tahoe, Nevada/California, Springer, November 2007.

[DPS02]   J. Diaz, J. Petit, and M. Serna. A survey of graph layout problems. *ACM Computing Surveys*, 34(3):313–356, 2002. DOI: 10.1145/568522.568523

[DS78]   D. Doo and M. Sabin. Behaviour of recursive division surfaces near extraordinary points. *Computer-Aided Design*, 10:356–360, September 1978. DOI: 10.1016/0010-4485(78)90111-2

[DSSD99]   X. Decoret, F. Sillion, G. Schaufler, and J. Dorsey. Multi-layered impostors for accelerated rendering. *Computer Graphics Forum*, 18(3):61–73, ISSN 1067-7055. September 1999. DOI: 10.1111/1467-8659.00328

[DSW07]   A. Dietrich, A. Stephens, and I. Wald. Exploring a boeing 777: Ray tracing large-scale cad data. *IEEE Computer Graphics and Applications*, 2007. DOI: 10.1109/MCG.2007.147

[Dur99]   F. Durand. *3D Visibility: Analytical Study and Applications*. Ph.D. thesis, Universite Joseph Fourier, Grenoble, France, 1999.

[DWS⁺97]   M. Duchaineau, M. Wolinsky, D.E. Sigeti, M.C. Miller, C. Aldrich, and M.B. Mineev-Weinstein. ROAMing Terrain: Real-time Optimally Adapting Meshes. In *Proceedings IEEE Visualization*, pp. 81–88, 1997. DOI: 10.1109/VISUAL.1997.663860

[EM99]   C. Erikson and D. Manocha. GAPS: General and automatic polygon simplification. In *Proceedings of ACM Symposium on Interactive 3D Graphics*, 1999. DOI: 10.1145/300523.300532

[EMB01]   C. Erikson, D. Manocha, and B. Baxter. HLODS for fast display of large static and dynamic environments. *Proceedings of ACM Symposium on Interactive 3D Graphics*, 2001. DOI: 10.1145/364338.364376

[FKST96]   T.A. Funkhouser, D. Khorramabadi, C.H. Sequin, and S. Teller. The ucb system for interactive visualization of large architectural models. *Presence*, 5(1):13–44, 1996.

[FLPR99]   M. Frigo, C.E. Leiserson, H. Prokop, and S. Ramachandran. Cache-oblivious algorithms. In *Foundations of Computer Science*, pp. 285–297, 1999. DOI: 10.1109/SFFCS.1999.814600

[FMP98]   L. De Floriani, P. Magillo, and E. Puppo. Efficient Implementation of Multi-Triangulations. In *Proceedings of IEEE Visualization 1998*, pp. 43–50, 1998. DOI: 10.1109/VISUAL.1998.745283

[FNB03]   M. Franquesa-Niubo and P. Brunet. Collision prediction using mktrees. *Proceedings CEIG*, pp. 217–232, 2003.

[FTW00]   P. Fishburn, P. Tetali, and P. Winkler. Optimal linear arrangement of a rectangular grid. *Discrete Mathematics*, 213(1):123–139, 2000. DOI: 10.1016/S0012-365X(99)00173-9

[GD98]   J.P. Grossman and W.J. Dally. Point Sample Rendering. In *Rendering Techniques 1998 (Proceedings of the Eurographics Workshop on Rendering)*, pp. 181–192, 1998.

[GGH02]   X. Gu, S.J. Gortler, and H. Hoppe. Geometry Images. In *ACM Transactions on Graphics (Proceedings of ACM SIGGRAPH)*, pp. 335–361, 2002. DOI: 10.1145/566570.566589

[GGSC96]   S.J. Gortler, R. Grzeszczuk, R. Szeliski, and M.F. Cohen. The lumigraph. In *Proceedings of SIGGRAPH 96*, Computer Graphics Proceedings, Annual Conference Series, pp. 43–54, August 1996. DOI: 10.1145/237170.237200

[GH97]   M. Garland and P.S. Heckbert. Surface simplification using quadric error metrics. In *Proceedings of SIGGRAPH 97*, Computer Graphics Proceedings, Annual Conference Series, pp. 209–216, August 1997. DOI: 10.1109/VISUAL.1998.745312

[GH98]   M. Garland and P.S. Heckbert. Simplifying surfaces with color and texture using quadric error metrics. In *IEEE Visualization '98*, pp. 263–270, October 1998. DOI: 10.1145/566570.566589

[GI99]   J. Gil and A. Itai. How to pack trees. *Journal of Algorithms*, 32(2):108–132, 1999. DOI: 10.1006/jagm.1999.1014

[Gib50]   J. Gibson. The perception of the visual world. Houghton Mifflin, Boston, MA, 1950.

[GJS76]   M. Garey, D. Johnson, and L. Stockmeyer. Some simplified np-complete graph problems. *Theoretical Computer Science 1*, pp. 237–267, 1976. DOI: 10.1016/0304-3975(76)90059-1

[GKM93]   N. Greene, M. Kass, and G. Miller. Hierarchical Z-Buffer Visibility. In *Computer Graphics (Proceedings of ACM SIGGRAPH)*, pp. 231–238, 1993. DOI: 10.1145/166117.166147

[GL96]   C. Gotsman and M. Lindenbaum. On the metric properties of discrete space-filling curves. *IEEE Transactions on Image Processing*, 5(5):794–797, 1996. DOI: 10.1109/83.499920

[GM04]   E. Gobbetti and F. Marton. Layered point clouds: A simple and efficient multiresolution structure for distributing and rendering gigantic point-sampled models. *Computers & Graphics*, 28(6):815–826, December 2004. DOI: 10.1016/j.cag.2004.08.010

[GM05]   E. Gobbetti and F. Marton. Far Voxels – a multiresolution framework for interactive rendering of huge complex 3d models on commodity graphics platforms. *ACM Transactions on Graphics*, 24(3):878–885, 2005. DOI: 10.1145/1073204.1073277

[GMC⁺06]   E. Gobbetti, F. Marton, P. Cignoni, M. Di Benedetto, and F. Ganovelli. C-BDAM – compressed batched dynamic adaptive meshes for terrain rendering. *Computer Graphics Forum*, 25(3):333–342, Proceedings Eurographics, 2006. DOI: 10.1111/j.1467-8659.2006.00952.x

[Gol81]   R. Goldstein. Defining the bounding edges of a synthavision solid model. In *18th Conference on Design Automation*, pp. 457–461, 1981.

[GP07]   M. Gross and H.-P. Pfister, Eds. *Point-based Graphics*. Elsevier Sciences Ltd., 2007.

[GSF99]   C. Gotsman, O. Sudarsky, and J.A. Fayman. Optimized occlusion culling using five-dimensional subdivision. *Computers & Graphics*, 23(5):645–654, October 1999.

DOI: 10.1016/S0097-8493(99)00088-6

[GSYM03]   N.K. Govindaraju, A. Sud, S.-E. Yoon, and D. Manocha. Interactive visibility culling in complex environments using occlusion-switches. In *2003 ACM Symposium on Interactive 3D Graphics*, pp. 103–112, April 2003. DOI: 10.1145/641480.641501

[GZ05]   M. Garland and Y. Zhou. Quadric-based simplification in any dimension. *ACM Transactions on Graphics*, 24(2):209–239, April 2005. DOI: 10.1145/1061347.1061350

[Hav00]   V. Havran. *Heuristic Ray Shooting Algorithms*. Ph.D. thesis, Department of Computer Science and Engineering, Faculty of Electrical Engineering, Czech Technical University in Prague, November 2000.

[HDD+93]   H. Hoppe, T. DeRose, T. Duchamp, J. McDonald, and W. Stuetzle. Mesh optimization. In J.T. Kajiya, Ed., *Computer Graphics (SIGGRAPH '93 Proceedings)*, volume 27, pp. 19–26, August 1993. DOI: 10.1145/166117.166119

[HHS06]   V. Havran, R. Herzog, and H.-. Seidel. On the fast construction of spatial data structures for ray tracing. In *Proceedings of IEEE Symposium on Interactive Ray Tracing 2006*, pp. 71–80, September 2006. DOI: 10.1109/RT.2006.280217

[HMC+97]   T. Hudson, D. Manocha, J. Cohem, M. Lin, K. Hoff, and H. Zhang. Accelerated occlusion culling using shadow frusta. In *Proceedings 13th Annu. ACM Sympos. Comput. Geom.*, pp. 1–10, 1997. DOI: 10.1145/262839.262847

[HMN05]   D. Haumont, O. Makinen, and S. Nirenstein. A low dimensional framework for exact polygon-to-polygon occlusion queries. In O. Deussen, A. Keller, K. Bala, P. Dutré, D.W. Fellner, and S.N. Spencer, Eds., *Rendering Techniques*, pp. 211–222. Eurographics Association, 2005.

[HMS06]   W. Hunt, W.R. Mark, and G. Stoll. Fast kd-tree construction with an adaptive error-bounded heuristic. In *2006 IEEE Symposium on Interactive Ray Tracing*. IEEE, September 2006. DOI: 10.1109/RT.2006.280218

[Hop96]   H. Hoppe. Progressive meshes. In H. Rushmeier, Ed., *SIGGRAPH 96 Conference Proceedings*, Annual Conference Series, pp. 99–108. ACM SIGGRAPH, Addison-Wesley, held in New Orleans, Louisiana, 04-09 August 1996. DOI: 10.1145/237170.237216

[Hop98]   H. Hoppe. Smooth view-dependent level-of-detail control and its applications to terrain rendering. In *IEEE Visualization '98 Conf.*, pp. 35–42, 1998. DOI: 10.1109/VISUAL.1998.745282

[Hop99a]   H. Hoppe. Optimization of mesh locality for transparent vertex caching. *ACM SIGGRAPH*, pp. 269–276, 1999. DOI: 10.1145/311535.311565

[Hop99b]   H.H. Hoppe. New quadric metric for simplifying meshes with appearance attributes. In *IEEE Visualization '99*, pp. 59–66, October 1999. DOI: 10.1109/VISUAL.1999.809869

[HPB05]   M. Heyer, S. Pfützer, and B. Brüderlin. Visualization Server for Very Large Virtual Reality Scenes. In *4. Paderborner Workshop Augmented & Virtual Reality in der Produktentstehung*, 2005.

[IG03]   M. Isenburg and S. Gumhold. Out-of-core compression for gigantic polygon meshes. In *ACM Trans. on Graphics (Proceedings of ACM SIGGRAPH)*, volume 22, pp. 935–942, 2003. DOI: 10.1145/882262.882366

[IL05]   M. Isenburg and P. Lindstrom. Streaming meshes. *IEEE Visualization*, pp. 231–238, 2005. DOI: 10.1109/VISUAL.2005.1532800

[ILGS03a]   M. Isenburg, P. Lindstrom, S. Gumhold, and J. Snoeyink. Large mesh simplification using processing sequences. *IEEE Visualization*, pp. 465–472, 2003. DOI: 10.1109/VISUAL.2003.1250408

[ILGS03b]   M. Isenburg, P. Lindstrom, S. Gumhold, and J. Snoeyink. Large Mesh Simplification using

Processing Sequences. In *Proceedings of IEEE Visualization 2003*, pp. 465–472, 2003.

[ISGM02]   W.V. Baxter III, A. Sud, N.K. Govindaraju, and D. Manocha. Gigawalk: Interactive walk-through of complex environments. In *Rendering Techniques 2002: 13th Eurographics Workshop on Rendering*, pp. 203–214, June 2002.

[JM92]   M. Juvan and B. Mohar. Optimal linear labelings and eigenvalues of graphs. *Discrete Applied Mathematics*, 36(2):153–168, 1992. DOI: 10.1016/0166-218X(92)90229-4

[JW02]   S. Jeschke and M. Wimmer. Textured depth meshes for realtime rendering of arbitrary scenes. In S. Gibson and P. Debevec, Eds., *Proceedings of the 13th Eurographics Workshop on Rendering (RENDERING TECHNIQUES-02)*, pp. 181–190, Aire-la-Ville, Switzerland, Eurographics Association, June 26-28 2002.

[JWS02]   S. Jeschke, M. Wimmer, and H. Schumann. Layered environment-map impostors for arbitrary scenes. In *Graphics Interface*, pp. 1–8, 2002.

[Kas04]   D. Kasik. Strategies for consistent image partitioning. *IEEE Multimedia*, 11(1):32–41, 2004. DOI: 10.1109/MMUL.2004.1261104

[KBG02]   Z. Karni, A. Bogomjakov, and C. Gotsman. Efficient compression and rendering of multi-resolution meshes. In *IEEE Visualization*, pp. 347–54, 2002. DOI: 10.1109/VISUAL.2002.1183794

[KCCO00]   V. Koltun, Y. Chrysanthou, and D. Cohen-Or. Virtual occluders: An efficient intermediate pvs representation. In *Rendering Techniques 2000: 11th Eurographics Workshop on Rendering*, pp. 59–70, June 2000.

[KCCO01]   V. Koltun, Y. Chrysanthou, and D. Cohen-Or. Hardware-accelerated from-region visibility using a dual ray space. In *Proceedings of the 12th Eurographics Workshop on Rendering Techniques*, pp. 205–216, London, UK, Springer-Verlag, 2001.

[KK98]   G. Karypis and V. Kumar. Multilevel k-way partitioning scheme for irregular graphs. *Journal of Parallel and Distributed Computing*, pp. 96–129, 1998. DOI: 10.1006/jpdc.1997.1404

[KKF99]   D. Kasik, C. Kimball, and J. Felt K. Frazier. A flexible approach to alliances of complex applications. *International Conference on Software Engineering*, 1999. DOI: 10.1109/ICSE.1999.840992

[KKM07]   A. Krishnamurthy, R. Khardekar, and S. McMains. Direct evaluation of nurbs curves and surfaces on the gpu. In *SPM '07: Proceedings of the 2007 ACM symposium on Solid and physical modeling*, pp. 329–334, New York, NY, ACM, 2007. DOI: 10.1145/1236246.1236293

[KLS96]   R. Klein, G. Liebich, and W. Straer. Mesh reduction with error control. In *IEEE Visualization '96*. IEEE, ISBN 0-89791-864-9, October 1996. DOI: 10.1109/VISUAL.1996.568124

[KMGL99]   S. Kumar, D. Manocha, W. Garret, and M. Lin. Hierarchical back-face computation. *Computer and Graphics*, 25(5):681–692, 1999. DOI: 10.1016/S0097-8493(99)00091-6

[KS00]   J.T. Klosowski and C.T. Silva. The Prioritized-Layered Projection Algorithm for Visible Set Estimation. In *IEEE Transaction on Visualization and Computer Graphics*, pp. 108–123, 2000. DOI: 10.1109/2945.856993

[KS01]   J.T. Klosowski and C.T. Silva. Efficient conservative visibility culling using the prioritized-layered projection algorithm. *IEEE Transactions on Visualization and Computer Graphics*, 7(4):365–379, 2001. DOI: 10.1109/2945.965350

[LB06]   C. Loop and J. Blinn. Real-time gpu rendering of piecewise algebraic surfaces. *ACM Transactions on Graphics*, 25(3):664–670, July 2006. DOI: 10.1145/1141911.1141939

[LCCO06]   A. Lerner, Y. Chrysanthou, and D. Cohen-Or. Efficient cells-and-portals partitioning: Research articles. *Comput. Animat. Virtual Worlds*, 17(1):21–40, 2006. DOI: 10.1002/cav.70

[LG95]   D.P. Luebke and C. Georges. Portals and mirrors: Simple, fast evaluation of potentially visible sets. In *Proceedings Symp. Interactive 3-D Graphics*, pp. 105–106, 1995.

[LH96]   M. Levoy and P. Hanrahan. Light field rendering. In *SIGGRAPH 96 Conference Proceedings*, pp. 31–42, 1996. DOI: 10.1145/237170.237199

[Lin00]   P. Lindstrom. Out-of-core simplification of large polygonal models. In *Proceedings of ACM SIGGRAPH 2000*, Computer Graphics Proceedings, Annual Conference Series, pp. 259–262, July 2000. DOI: 10.1145/344779.344912

[Lin03]   P. Lindstrom. Out-of-core construction and visualization of multiresolution surfaces. In *ACM 2003 Symposium on Interactive 3D Graphics*, pp. 93–102, 2003. DOI: 10.1145/641480.641500

[Lip80]   A. Lippman. Movie-maps: An application of the optical videodisc to computer graphics. *Computer Graphics (SIGGRAPH '80 Proceedings)*, 14(3):32–42, July 1980. DOI: 10.1145/965105.807465

[LP01]   P. Lindstrom and V. Pascucci. Visualization of large terrains made easy. *IEEE Visualization*, pp. 363–370, 2001. DOI: 10.1109/VISUAL.2001.964533

[LRC+02]   D. Luebke, M. Reddy, J. Cohen, A. Varshney, B. Watson, and R. Huebner. *Level of Detail for 3D Graphics: Applications and Theory.* Morgan Kaufmann, 2002.

[LSCO03]   T. Leyvand, O. Sorkine, and D. Cohen-Or. Ray space factorization for from-region visibility. *ACM Transactions on Graphics*, 22(3):595–604, July 2003. DOI: 10.1145/882262.882313

[Lue01]   D.P. Luebke. A Developer's Survey of Polygonal Simplification Algorithms. *IEEE Computer Graphics and Applications*, 21(3):24–35, 2001. DOI: 10.1109/38.920624

[LV03]   P. Lyman and H. Varian. How much information? http://www2.sims.berkeley.edu/research/projects/how-much-info/, 2003.

[LW85]   M. Levoy and T. Whitted. The Use of Points as a Display Primitive. Technical Report TR 85-022, University of North Carolina at Chapel Hill, 1985.

[LYTM06]   C. Lauterbach, S.-E. Yoon, D. Tuft, and D. Manocha. Rt-deform: Interactive ray tracing of dynamic scenes using bvhs. In *Proceedings of IEEE Symposium on Interactive Ray Tracing 2006*, September 2006. DOI: 10.1109/RT.2006.280213

[MAM05]   F. Mora, L. Aveneau, and M. Mériaux. Coherent and exact polygon-to-polygon visibility. In *Proceedings WSCG*, pp. 87–94, 2005.

[MB95]   L. McMillan and G. Bishop. Plenoptic Modeling: An Image-Based Rendering System. In *ACM Computer Graphics (Proceedings of ACM SIGGRAPH)*, pp. 39–46, 1995.

[MBW06]   O. Mattausch, J. Bittner, and M. Wimmer. Adaptive visibility-driven view cell construction. In *Rendering Techniques 2006: 17th Eurographics Workshop on Rendering*, pp. 195–206, June 2006.

[MBWW07]   O. Mattausch, J. Bittner, P. Wonka, and M. Wimmer. Optimized subdivisions for preprocessed visibility. In *Graphics Interface 2007*, pp. 335–342, May 2007. DOI: 10.1145/1268517.1268571

[Moo65]   G. Moore. Cramming more components onto integrated circuits. *Electronics Magazine*, 38(8):114–117, 1965. DOI: 10.1109/JPROC.1998.658762

[Moo91]   A. Moore. *A Tutorial on kd-trees.* Ph.D. thesis, University of Cambridge, 1991.

[MP80]   D. Muradyan and T. Piliposyan. Minimal numberings of vertices of a rectangular lattice. In *Akad. Nauk. Arimjan*, pp. 21–27, 1980.

[NB04]   S. Nirenstein and E. Blake. Hardware accelerated visibility preprocessing using adaptive sampling. In *Rendering Techniques 2004: 15th Eurographics Workshop on Rendering*, pp. 207–216, June 2004.

[NBG02]   S. Nirenstein, E. Blake, and J. Gain. Exact from-region visibility culling. In *EGRW '02: Proceedings of the 13th Eurographics workshop on Rendering*, pp. 191–202, Aire-la-Ville, Switzerland, Switzerland, Eurographics Association, 2002.

[ND05]   M.A. Sabin N. Dodgson, M.S. Floater, Eds. *Advances in Multiresolution for Geometric Modelling*. Springer, 2005.

[NFLYCO99]   B. Nadler, G. Fibich, S. Lev-Yehudi, and D. Cohen-Or. A qualitative and quantitative visibility analysis in urban scenes. *Computers & Graphics*, 23(5):655–666, October 1999. DOI: 10.1016/S0097-8493(99)00089-8

[NRS97]   R. Niedermeier, K. Reinhardt, and P. Sanders. Towards optimal locality in mesh-indexings. In *Fundamentals of Computation Theory*, pp. 364–375, 1997. DOI: 10.1007/BFb0036198

[NRS02]   R. Niedermeier, K. Reinhardt, and P. Sanders. Towards optimal locality in mesh-indexings. *Discrete Applied Mathematics*, 117(1):211–237, 2002. DOI: 10.1016/S0166-218X(00)00326-7

[OBM00]   M.M. Oliveira, G. Bishop, and D. McAllister. Relief Texture Mapping. In *ACM Computer Graphics (Proceedings of ACM SIGGRAPH)*, pp. 359–368, 2000. DOI: 10.1145/344779.344947

[oM95]   National Library of Medicine. The visible human project. NL Medicine, 1995.

[PAC+97]   D. Patterson, T. Anderson, N. Cardwell, R. Fromm, K. Keaton, C. Kazyrakis, R. Thomas, and K. Yellick. A case for intelligent ram. *IEEE Micro.*, 34–44, 1997. DOI: 10.1109/40.592312

[PCM07]   Flicker fusion rate. *PCMag.Com Encyclopedia Terms*, 2007.

[PF01]   V. Pascucci and R.J. Frank. Global static indexing for real-time exploration of very large regular grids. In *Supercomputing*, 363–370, 2001. DOI: 10.1145/582034.582036

[PG07]   R. Pajarola and E. Gobbetti. Survey on semi-regular multiresolution models for interactive terrain rendering. *The Visual Computer*, 23(8):583–605, 2007. DOI: 10.1007/s00371-007-0163-2

[PGSS06]   S. Popov, J. Günther, H.-P. Seidel, and P. Slusallek. Experiences with streaming construction of SAH KD-trees. In *Proceedings of the 2006 IEEE Symposium on Interactive Ray Tracing*, pp. 89–94, September 2006. DOI: 10.1109/RT.2006.280219

[POC05]   F. Policarpo, M.M. Oliveira, and J.L.D. Comba. Real-time relief mapping on arbitrary polygonal surfaces. *ACM Trans. Graph*, 24(3):935, 2005. DOI: 10.1145/1073204.1073292

[Pri00]   C. Prince. Progressive meshes for large models of arbitrary topology. Master's thesis, Department of Computer Science and Engineering, University of Washington, Seattle, August 2000.

[RB93]   J. Rossignac and P. Borrel. Multi-resolution 3D approximation for rendering complex scenes. In *Second Conference on Geometric Modelling in Computer Graphics*, pp. 453–465, Genova, Italy, June 1993.

[RL00a]   S. Rusinkiewicz and M. Levoy. Qsplat: A multiresolution point rendering system for large meshes. *Proceedings of ACM SIGGRAPH*, pp. 343–352, 2000.

[RL00b]   S. Rusinkiewicz and M. Levoy. Qsplat: A multiresolution point rendering system for large meshes. In *Proceedings of ACM SIGGRAPH 2000*, Computer Graphics Proceedings, Annual Conference Series, pp. 343–352, July 2000.

[RL00c]   S. Rusinkiewicz and M. Levoy. QSplat: A Multiresolution Point Rendering System for Large Meshes. In *Computer Graphics (Proceedings of ACM SIGGRAPH)*, pp. 343–352, 2000.

[RR96]   R. Ronfard and J. Rossignac. Full-range approximation of triangulated polyhedra. *Computer Graphics Forum (Eurographics'96 Proceedings)*, 15(3):67–76, 1996.

DOI: 10.1111/1467-8659.1530067

[RSH05]   A. Reshetov, A. Soupikov, and J. Hurley. Multi-Level Ray Tracing Algorithm. In *ACM Transaction of Graphics (Proceedings of ACM SIGGRAPH)*, pp. 1176–1185, 2005. DOI: 10.1145/1073204.1073329

[RW94]   C. Ruemmler and J. Wilkes. An introduction to disk drive modeling. *IEEE Computer*, 1994.

[Sag94]   H. Sagan. *Space-Filling Curves*. Springer-Verlag, 1994.

[Sam06]   H. Samet, Ed. *Foundations of Multidimensional and Metric Data Structures*. Morgan Kaufmann, 2006.

[SC97]   M. Slater and Y. Chrysanthou. View volume culling using a probabilistic caching scheme. In *Proceedings of the ACM Symposium on Virtual Reality Software and Technology*, pp. 71–78, 1997. DOI: 10.1145/261135.261150

[SCC+02]   C. Silva, Y.-J. Chiang, W. Correa, J. El-Sana, and P. Lindstrom. Out-of-core algorithms for scientific visualization and computer graphics. In *IEEE Visualization Course Notes*, 2002.

[SCD02]   S. Sen, S. Chatterjee, and N. Dumir. Towards a theory of cache-efficient algorithms. *Journal of the ACM*, 49:828–858, 2002. DOI: 10.1145/602220.602225

[SD01]   M. Stamminger and G. Drettakis. Interactive Sampling and Rendering for Complex and Procedural Geometry. In *Proceedings of the Eurographics Workshop on Rendering Techniques*, pp. 151–162, 2001.

[SDB97]   F. Sillion, G. Drettakis, and B. Bodelet. Efficient Impostor Manipulation for Real-Time Visualization of Urban Scenery. In *Computer Graphics Forum (Proceedings of Eurographics)*, pp. 207–218, 1997. DOI: 10.1111/1467-8659.00158

[SG01]   E. Shaffer and M. Garland. Efficient adaptive simplification of massive meshes. In *IEEE Visualization 2001*, pp. 127–134, October 2001. DOI: 10.1109/VISUAL.2001.964503

[SGHS98]   J. Shade, S. Gortler, L.-W. He, and R. Szeliski. Layered Depth Images. In *Computer Graphics (Proceedings of ACM SIGGRAPH)*, pp. 231–242, 1998. DOI: 10.1145/280814.280882

[SGwHS98]   J. Shade, S.J. Gortler, L.-W. He, and R. Szeliski. Layered depth images. In *Proceedings of SIGGRAPH 98*, Computer Graphics Proceedings, Annual Conference Series, pp. 231–242, July 1998.

[SJDS00]   G. Schaufler, J.Dorsey, X. Decoret, and F.X. Sillion. Conservative volumetric visibility with occluder fusion. In *SIGGRAPH 00 Conference Proceedings*, pp. 229–238, 2000. DOI: 10.1145/344779.344886

[SMS+07]   M. Shevtsov, A. Soupikov, and A. Kapustin. Highly parallel fast kd-tree construction for interactive ray tracing of dynamic scenes. *Computer Graphics Forum*, 26(3):395–404, September 2007. DOI: 10.1111/j.1467-8659.2007.01062.x

[Ste97]   A.J. Stewart. Hierarchical visibility in terrains. In *Eurographics Rendering Workshop 1997*, pp. 217–228, June 1997.

[Ste07]   J. Stevens. Concepts and concerns related to the visualization of complex automotive data. *IEEE Computer Graphics and Applications*, 2007. DOI: 10.1109/MCG.2007.161

[Str74]   W. Strasser. *Schnelle Kurven- und Flaechendarstellung auf graphischen Sichtgeraeten*. Ph.D. thesis, TU Berlin, 1974. DOI: 10.1109/MCG.2007.161

[Stu99]   W. Stuerzlinger. Imaging all visible surfaces. In *Graphics Interface '99*, pp. 115–122, June 1999.

[Sut63]   I.E. Sutherland. Sketchpad: A man-machine graphical communication system. *SJCC*, 1963.

[SWBG06]   C. Sigg, T. Weyrich, M. Botsch, and M. Gross. Gpu-based ray-casting of quadratic surfaces. In *Symposium on Point - Based Graphics 2006*, pp. 59–66, July 2006.

[SWS07]   K. Sun, G. Watson, and C. Seeling. Shader algorithm for the interactive, stereoscopic visualization of crash worthiness simulations. *IEEE Computer Graphics and Applications*, 2007.

[SZL92]   W.J. Schroeder, J.A. Zarge, and W.E. Lorensen. Decimation of Triangle Meshes. In *ACM Computer Graphics (Proceedings of ACM SIGGRAPH)*, pp. 65–70, 1992. DOI: 10.1145/142920.134010

[TC05]   J. Thomas and K. Cook. Illuminating the path: The research and development agenda for visual analytics. *IEEE Press*, 2005.

[TCM06]   M. Tarini, P. Cignoni, and C. Montani. Ambient occlusion and edge cueing for enhancing real time molecular visualization. *IEEE Transactions on Visualization and Computer Graphics*, 12(5):1237–1244, September/October 2006. DOI: 10.1109/TVCG.2006.115

[Tel92]   S.J. Teller. Computing the antipenumbra of an area light source. In *Computer Graphics (Proceedings of SIGGRAPH 92)*, pp. 139–148, July 1992. DOI: 10.1145/142920.134029

[TS91]   S.J. Teller and C.H. Sequin. Visibility preprocessing for interactive walk-throughs. *Computer Graphics (SIGGRAPH 91 Proceedings)*, 25(4):61–69, July 1991. DOI: 10.1145/127719.122725

[VdMG91]   L. Velho and J. de Miranda Gomes. Digital halftoning with space filling curves. In *ACM SIGGRAPH*, pp. 81–90, 1991. DOI: 10.1145/127719.122727

[vdPS99]   M. van de Panne and J. Stewart. Efficient compression techniques for precomputed visibility. In *Eurographics Rendering Workshop 1999*, June 1999.

[vEB77]   P. van E. Boas. Preserving order in a forest in less than logarithmic time and linear space. *Inf. Process. Lett.*, 1977. DOI: 10.1016/0020-0190(77)90031-X

[Vit01]   J. Vitter. External memory algorithms and data structures: Dealing with massive data. *ACM Computing Surveys*, pp. 209–271, 2001. DOI: 10.1145/384192.384193

[vKvOB+97]   M. van Kreveld, R. van Oostrum, C. Bajaj, V. Pascucci, and D.R. Schikore. Contour trees and small seed sets for isosurface traversal. In *Symp. on Computational Geometry*, 1997. DOI: 10.1145/262839.269238

[Wal04]   I. Wald. *Realtime Ray Tracing and Interactive Global Illumination*. Ph.D. thesis, Computer Graphics Group, Saarland University, 2004.

[Wal07]   I. Wald. On fast construction of SAH-based bounding volume hierarchies. In *Proceedings of the 2007 Eurographics/IEEE Symposium on Interactive Ray Tracing*, 2007. DOI: 10.1109/RT.2007.4342588

[WBP98]   Y. Wang, H. Bao, and Q. Peng. Accelerated walkthroughs of virtual environments based on visibility preprocessing and simplification. *Computer Graphics Forum*, 17(3), 1998. DOI: 10.1111/1467-8659.00266

[WBS07]   I. Wald, S. Boulos, and P. Shirley. Ray tracing deformable scenes using dynamic bounding volume hierarchies. *ACM Transactions on Graphics*, 26(1):6.1–6.10, 2007. DOI: 10.1145/1276377.1276490

[WDS04]   I. Wald, A. Dietrich, and P. Slusallek. An Interactive Out-of-Core Rendering Framework for Visualizing Massively Complex Models. In *Proceedings of the Eurographics Symposium on Rendering*, 2004. DOI: 10.1145/1198555.1198756

[WFH+07]   T. Weyrich, C. Flaig, S. Heinzle, S. Mall, T. Aila, K. Rohrer, D. Fasnacht, N. Felber, S. Oetiker, H. Kaeslin, M. Botsch, and M. Gross. A Hardware Architecture for Surface Splatting. In *ACM Transactions on Graphics (Proceedings of ACM SIGGRAPH)*, p. 90, 2007. DOI: 10.1145/1276377.1276490

[WFP+01]   M. Wand, M. Fischer, I. Peter, F. Meyer, and W. Straßer. The Randomized z-Buffer Algorithm: Interactive Rendering of Highly Complex Scenes. In *Computer Graphics (Proceedings of ACM SIGGRAPH)*, pp. 361–370, 2001.

[WH06]   I. Wald and V. Havran. On building fast kd-trees for ray tracing, and on doing that in o(n log n). In *Proceedings of IEEE Symposium on Interactive Ray Tracing 2006*, pp. 61–69, 2006. DOI: 10.1109/RT.2006.280216

[Wie02]   J.-M. Wierum. Logarithmic path-length in space-filling curves. In *14th Canadian Conference on Computational Geometry*, pp. 22–26, 2002.

[Wik07]   Flicker fusion threshold article. *Wikipedia*, 2007.

[WK03]   J. Wu and L. Kobbelt. A stream algorithm for the decimation of massive meshes. In *Proceedings Graphics Interface*, pp. 185–192, 2003.

[WK06]   C. Wächter and A. Keller. Instant ray tracing: The bounding interval hierarchy. In *Proceedings of the Eurographics Symposium on Rendering*, pp. 139–149, 2006.

[WLML99]   A. Wilson, E. Larsen, D. Manocha, and M.C. Lin. Partitioning and handling massive models for interactive collision detection. *Computer Graphics Forum (Proceedings of Eurographics)*, 18(3):319–329, 1999. DOI: 10.1111/1467-8659.00352

[WM03]   A. Wilson and D. Manocha. Simplifying Complex Environments Using Incremental Textured Depth Meshes. In *ACM Transactions on Graphics (Proceedings of ACM SIGGRPAH)*, pp. 678–688, 2003. DOI: 10.1145/882262.882325

[WMS06]   S. Woop, G. Marmitt, and P. Slusallek. B-KD Trees for Hardware Accelerated Ray Tracing of Dynamic Scenes. In *Proceedings of Graphics Hardware*, 2006.

[WPS⁺03]   I. Wald, T.J. Purcell, J. Schmittler, C. Benthin, and P. Slusallek. Realtime Ray Tracing and its use for Interactive Global Illumination. In *Eurographics 2003 State of the Art Reports*, 2003.

[WSBW01]   I. Wald, P. Slusallek, C. Benthin, and M. Wagner. Interactive rendering with coherent ray tracing. *Computer Graphics Forum*, 20(3):153–164, 2001. DOI: 10.1111/1467-8659.00508

[WSS05]   S. Woop, J. Schmittler, and P. Slusallek. RPU: A Programmable Ray Processing Unit for Realtime Ray Tracing. In *ACM Transactions on Graphics (Proceedings of ACM SIGGRAPH)*, pp. 434–444, 2005. DOI: 10.1145/1073204.1073211

[WTL⁺04]   X. Wang, X. Tong, S. Lin, S. Hu, B. Guo, and H.-Y. Shum. Generalized displacement maps. In D. Fellner and S. Spencer, Eds., *Proceedings of the 2004 Eurographics Symposium on Rendering*, pp. 227–234. Eurographics Association, June 2004.

[WWS00]   P. Wonka, M. Wimmer, and D. Schmalstieg. Visibility preprocessing with occluder fusion for urban walkthroughs. In *11th Eurographics Workshop on Rendering*, pp. 71–82, 2000.

[WWS01a]   M. Wimmer, P. Wonka, and F. Sillion. Point-based impostors for real-time visualization, May 29, 2001.

[WWS01b]   P. Wonka, M. Wimmer, and F.X. Sillion. Instant visibility. *Computer Graphics Forum*, 20(3):411–421, 2001. DOI: 10.1111/1467-8659.00534

[WWT⁺03]   L. Wang, X. Wang, X. Tong, S. Lin, S.-M. Hu, B. Guo, and H.-Y. Shum. View-Dependent Displacement Mapping. In *ACM Transactions on Graphics (Proceedings of ACM SIGGRAPH)*, pp. 334–339, 2003. DOI: 10.1145/882262.882272

[WWZ⁺06]   P. Wonka, M. Wimmer, K. Zhou, S. Maierhofer, G. Hesina, and A. Reshetov. Guided visibility sampling. *ACM Transactions on Graphics*, 25(3):494–502, July 2006. DOI: 10.1145/1141911.1141914

[YCM07]   S. Yoon, S. Curtis, and D. Manocha. Ray tracing dynamic scenes using selective restructuring. *Proceedings of Eurographics Symposium on Rendering*, 2007.

[YL06]   S.-E. Yoon and P. Lindstrom. Mesh layouts for block-based caches. *IEEE Transactions on Visualization and Computer Graphics (Proceedings Visualization)*, 12(5), 2006. DOI: 10.1109/TVCG.2006.162

[YLM06]   S.-E. Yoon, C. Lauterbach, and D.   Manocha. R-LODs: Fast LOD-Based

Ray Tracing of Massive Models. *The Visual Computer*, 22(9-11):772–784, 2006. DOI: 10.1007/s00371-006-0062-y

[YLPM05]   S.-E. Yoon, P. Lindstrom, V. Pascucci, and D. Manocha. Cache-Oblivious Mesh Layouts. *Proceedings of ACM SIGGRAPH*, 2005. DOI: 10.1145/1186822.1073278

[YM06]   S.-E. Yoon and D. Manocha. Cache-efficient layouts of bounding volume hierarchies. *Computer Graphics Forum (Eurographics)*, 25:507–516, 2006. DOI: 10.1111/j.1467-8659.2006.00970.x

[Yoo05]   S.-E. Yoon. *Interactive Visualization and Collision Detection using Dynamic Simplification and Cache-Coherent Layouts*. Ph.D. thesis, University of North Carolina at Chapel Hill, 2005.

[YSGM04]   S.-E. Yoon, B. Salomon, R. Gayle, and D. Manocha. Quick-VDR: Interactive View-dependent Rendering of Massive Models. *IEEE Visualization*, pp. 131–138, 2004.

[YSGM05]   S.-E. Yoon, B. Salomon, R. Gayle, and D. Manocha. Quick-VDR: Out-of-Core View-Dependent Rendering of Gigantic Models. *IEEE Transactions on Visualization and Computer Graphics*, pp. 369–382, 2005. DOI: 10.1109/TVCG.2005.64

[YSLM04]   S. Yoon, B. Salomon, M. C. Lin, and D. Manocha. Fast collision detection between massive models using dynamic simplification. *Eurographics Symposium on Geometry Processing*, pp. 136–146, 2004. DOI: 10.1145/1057432.1057450

[ZMHH97]   H. Zhang, D. Manocha, T. Hudson, and K. Hoff. Visibility culling using hierarchical occlusion maps. *Proceedings of ACM SIGGRAPH*, 1997. DOI: 10.1145/258734.258781

[ZT02]   E. Zhang and G. Turk. Visibility-guided simplification. In *Proceedings IEEE Visualization*, pp. 267–274, 2002. DOI: 10.1109/VISUAL.2002.1183784

Biographies

Enrico Gobbetti is the director of the Advanced Computing and Communications Program and of the Visual Computing group at the CRS4 resarch center in Italy. He holds an Engineering degree (1989) and a Ph.D. degree (1993) in Computer Science from the Swiss Federal Institute of Technology in Lausanne (EPFL). Prior to joining CRS4, he held research and teaching positions at the Swiss Federal Institute of Technology in Lausanne, the University of Maryland Baltimore County, and the NASA Center of Excellence in Space Data and Information Sciences. At CRS4, Enrico developed and managed a research program supported through industrial and government grants. His research spans many areas of computer graphics and is widely published in major journals and conferences. He regularly serves as program committee member or reviewer for international conferences and journals and is currently Associate Editor of Computer Graphics Forum. Technologies developed by his group have found practical use in as diverse real-world applications as internet geoviewing, scientific data analysis, and surgical training.

Dave Kasik, Boeing Senior Technical Fellow, is responsible for visualization and interactive techniques across the enterprise. His research interests include innovative combinations of basic 3D graphics and user interface technologies and increasing awareness of the impact of visualization technology inside and outside Boeing. Dave has a BA in Quantitative Studies from the Johns Hopkins University and an MS in Computer Science from the University of Colorado. He is an ACM Distinguished Scientist and a member of IEEE, ACM SIGGRAPH, and ACM SIGCHI. He is a member of the editorial board for IEEE Computer Graphics and Applications.

Dinesh Manocha is currently a Phi Delta Theta/Mason Distinguished Professor of Computer Science at the University of North Carolina at Chapel Hill. He received his Ph.D. in Computer Science at the University of California at Berkeley 1992. He received Junior Faculty Award in 1992, Alfred P. Sloan Fellowship and NSF Career Award in 1995, Office of Naval Research Young Investigator Award in 1996, Honda Research Initiation Award in 1997, and Hettleman Prize for Scholarly Achievements at UNC Chapel Hill in 1998. He has also received eight best paper & panel awards at top conferences in graphics, modeling, simulation and visualization. He has been working on technologies for displaying massive models for more than 10 years. Many of the technologies developed by his group on collison detection, GPU-based algorithms and large model rendering have been widely used. He has published more than 210 papers in leading conferences and journals on computer graphics, geometric modeling, robotics, virtual environments and computational geometry. He has also served as a program committee member for more than 50 leading conferences in these areas and also served in the editorial board of many journals.

Sung-Eui Yoon is currently an assistant professor at Korea Advanced Institute of Science and Technology (KAIST). He received the B.S. and M.S. degrees in computer science from Seoul National University in 1999 and 2001 respectively. He received his Ph.D. degree in computer science from the University of North Carolina at Chapel Hill in 2005. He was a postdoctoral scholar at Lawrence Livermore National Laboratory. His research interests include visualization, interactive rendering, geometric problems, and cache-coherent algorithms and layouts.

Index